Freely Suspended Liquid Crystalline Films

Freely Suspended Liquid Crystalline Films

A. A. Sonin
Centre d'Etudes Atomiques de Saclay, France
and
Institute of Crystallography, Russian Academy of Sciences

with a Foreword by

Noel Clark
University of Colorado, USA

JOHN WILEY & SONS
Chichester · New York · Weinheim · Brisbane · Singapore · Toronto

Copyright © 1998 John Wiley & Sons Ltd,
Baffins Lane, Chichester,
West Sussex PO19 1UD, England

National   01243 779777
International   (+44) 1243 779777
e-mail (for orders and customer service enquiries): cs-books@wiley.co.uk
Visit our Home Page on http://www.wiley.co.uk
or http://www.wiley.com

Other Wiley Editorial Offices

John Wiley & Sons, Inc., 605 Third Avenue,
New York, NY 10158-0012, USA

WILEY-VCH Verlag GmbH, Pappelallee 3,
D-69469 Weinheim, Germany

Jacaranda Wiley Ltd, 33 Park Road, Milton,
Queensland 4064, Australia

John Wiley & Sons (Asia) Pte Ltd, Clementi Loop #02-01,
Jin Xing Distripark, Singapore 129809

John Wiley & Sons (Canada) Ltd, 22 Worcester Road,
Rexdale, Ontario M9W 1L1, Canada

Library of Congress Cataloging-in-Publication Data

Sonin, A. A.
   Freely suspended liquid crystalline films / A.A. Sonin.
      p.   cm.
   Includes bibliographical references and index.
   ISBN 0-471-97155-3 (alk. paper)
   1. Liquid crystal films.   I. Title.
   QD923.S658   1998   98-6336
   530.4'29—dc21   CIP

British Library Cataloguing in Publication Data

A catalogue record for this book is available from the British Library

ISBN 0 471 97155 3

Typeset by Techset Composition Limited, Salisbury, England
Printed and bound in Great Britain by Bookcraft (Bath) Ltd
This book is printed on acid-free paper responsibly manufactured from sustainable forestry,
in which at least two trees are planted for each one used for paper production

To the memory of Valeria

Contents

Foreword

Freely suspended liquid films and bubbles have proved to be a topic of continuing human interest, from the particular fascination of children of all ages, to the scientific investigation of recent centuries. Indeed, freely suspended soap films of surfactant and water have attracted the attention of leading physicists and chemists in every generation since Newton, and have played an essential role in the evolution of our understanding of the interfacial properties of fluids and biological membranes, and in the development of surfactants.

Freely suspended films have been a wellspring of new liquid crystal science since the discovery in the 1970s that films as thin as a single smectic layer could be made of single-molecular components having bulk liquid crystal phases. Such films are the thinnest known stable condensed phase structures and have the largest surface-to-volume ratio, making them ideal for studies of two-dimensional phase transitions and of fluctuation and interface phenomena. The study of these films has been crucial to the current understanding of liquid crystal phase behavior, and has addressed important issues in condensed matter physics, providing unique experimental conditions for the study of phase transitions in two dimensions. They are the only system in which the hexatic has unambiguously been identified as a phase of matter, and the only physical system in which fluctuations of a 2D XY system and Kosterlitz Thouless phase transition have been visualized and 2D XY quasi long range order observed. Films have enabled the precise determination of smectic-layer electron density and positional fluctuations, and have been used to show that the interlayer interactions in tilted smectics are predominantly nearest neighbor.

In this monograph Professor Sonin provides a comprehensive overview of both multicomponent surfactant-based films and single component thermotropic smectic liquid crystal films. He deals with basic issues of film stability, equilibrium structure, and thinning in the context of surfactant-based films, as well as the phase transitions, orientational behavior and defect structures of thermotropic films. This book will serve as a valuable resource for students and researchers in complex fluid science, as well as for those who desire a quick introduction to the beautiful world of freely suspended liquid crystal films.

Noel Clark
Condensed Matter Laboratory
University of Colorado

Preface

Liquid crystals (or mesophases)—the state of matter intermediate between a solid crystal and an amorphous liquid—attract a lot of attention from scientists and engineers due to their interesting physical properties and practical applications.

This monograph concerns the physics of freely suspended (or free-standing) liquid crystalline (FSLC) films, i.e. anisotropic films stretched over a solid frame and contacting on both sides with air (or, more rarely, with some other gaseous phase or a vacuum).

FSLC films with a thickness of more than several hundred Å can be obtained using only mesomorphic liquids. However, for preparation of thinner FSLC films (thickness less than several hundred Å) a great variety of amorphous fluids can be used; orientational molecular ordering in this case can be induced due the action of the surface forces.

FSLC films are very interesting objects to study, first of all, due to the pronounced surface effects. Indeed, because of the presence of the long-range orientational molecular order in such films, the influence of the bounding interfaces is considerable, even for quite thick films (with a thickness of the order of several tens of μm), while the properties of films with such thicknesses, prepared for example from liquid helium or metals, are indistinguishable from those of massive bulk samples.

In addition, the FSLC films can be regarded as 'pure' systems for studying surface phenomena in liquid crystals. Indeed, for investigations and practical applications, mesophases are generally placed in flat glass or quartz capillaries, i.e. liquid crystal is in contact with solid walls. As a consequence, the surface forces depend both on the structure of the solid boundaries and on the mesophase. These walls always possess some geometrical inhomogenities (roughness), which also influence the surface mesophase layers. All this complicates very much the investigation of the surface properties of liquid crystals.

For a FSLC film, we have free bounding surfaces. This simplifies considerably the situation: now the surface forces are determined mainly by the structure of the liquid medium. In addition, the film interfaces can often (but not always) be considered as atomically smooth, and its roughness, therefore, is negligible.

Moreover, experimentally, it is much easier to obtain very thin (with a thickness of less than 1000 Å) free-standing liquid films (and FSLC films, in particular), than thin liquid films confined by solid substrates.

The second point is that FSLC films can provide interesting examples of objects with reduced dimensionality and hence with unusual physical properties. Indeed, free-standing film can often be considered as a two-dimensional system, since the size of the periphery is generally much greater than the thickness. In addition, the presence of optical anisotropy in FSLC films gives the possibility of easy visual observations (in a polarisation microscope) of many complex phenomena, such as two-dimensional topological defects, two-dimensional hydrodynamics, etc.

FSLC films are attractive not only from the point of view of pure science; they are also of considerable practical interest. Indeed, the problems of stability, thinning and rupture of these films are important in the oil, chemical and cosmetics industries (e.g. for oil extraction, for production of various foams and detergents, etc.). The problems of molecular organisation, phase transitions, elasticity, etc., for these films are of interest for medical and biological applications (e.g. the structure of biomembranes, molecular mechanisms of cancer diseases, drug transport through cell membranes, etc.).

In this book, some basic notions concerning liquid crystalline phases are firstly given. Then, some principal macroscopic physical parameters characterising FSLC films are introduced. After that, experimental techniques for preparation of these films, and measurements of their thickness, for investigations of their physical properties and structure, are reviewed. Further on, the problems of film stability, thinning and rupture are discussed. Then some physical phenomena in these films, such as orientational and phase transitions, and field-induced effects, are analysed. Further still, several examples of molecular structures of thermodynamically stable FSLC films are given, and the topological defects in some FSLC films are considered. Finally, some applications of these films in practice and in some other (non-physical) sciences are outlined.

Quite a large number of books and review articles exist that are dedicated to the physics and physical chemistry of liquid (mainly soap and foam) freely suspended films. Nevertheless, this modest work is the first monograph dedicated exclusively to FSLC films, and the author hopes that it will stimulate interest in these objects and research in this very promising area. It should be noted, however, that some aspects of the discussed topic have already been considered in recent review papers [for example *Thin Liquid Films*, 1988, Surfactant Science Ser. (ed. I.B. Ivanov), Marcel Dekker, New York; vol. 29; P. Pieranski *et al.*, 1993, *Physica A*, **194**, 364; Ch. Bahr, 1994, *Int. J. Mod. Phys. B*, **8**, 3051; T. Stoebe and C. C. Huang, 1995, *Int. J. Mod. Phys. B*, **9**, 2285], in books on soap films and foams [e.g. P.M. Kruglyakov and Yu.G. Rovin, 1978, *Physical Chemistry of Black Hydrocarbon Films. Biomolecular Lipid Membranes*, Nauka, Moscow (in Russian); S.S. Duhin, N.N. Rulyev and D.S. Dimitrov, 1986, *Coagulation and Dynamics of Thin Films*, Naukova Dumka, Kiev (in Russian); P.M. Kruglyakov and D.R. Ekserova, 1990, *Foam and Foam Films*,

Khimia, Moscow (in Russian)], and in books on liquid crystals [e.g. P.S. Pershan, 1988, *Structure of Liquid Crystal Phases*, World Scientific, Singapore; E.I. Kats and V.V. Lebedev, 1994, *Fluctuational Effects in the Dynamics of Liquid Crystals*, Springer, New York (Russian version: 1988, Nauka, Moscow); A.A. Sonin, 1995, *The Surface Physics of Liquid Crystals*, OPA–Gordon and Breach, Amsterdam].

I wish to express my gratitude to L.M. Blinov and E.I. Kats, in co-operation with whom I have studied surface phenomena in liquid crystals for about 10 years at the Institute of Crystallography, Russian Academy of Sciences (Moscow). I also appreciate the collaboration with my foreign colleagues: D. Langevin, A. Bonfillon, J. Bechhoefer, B. Frisken, A. Yethiraj, T. Palermo and A. Lubek, with whom I have worked for the last few years, in France and Canada, on problems connected with free-standing films. I am deeply grateful to A.S. Sonin for his valuable advice and help in the bibliographical searches.

Finally, my sincere gratitude goes to Myriam Jiménez Sierra, for her patience with me during quite a long period of work on this book in Paris.

Paris, November 1997   A.A.S.

1

Introduction

1.1 FREE-STANDING FILMS—HISTORY

People have observed free-standing liquid films since ancient times in the form of soap bubbles. Indeed, soap was invented quite a long time ago. In his book *Historia Naturalis* the ancient-Roman philosopher Gaius Plinius already mentioned soap films. He wrote that in Gaul and Germany soap was prepared from animal fat and alkaline solution. The corrosive properties of the latter material were increased by adding lime. Even solid and liquid soaps were already distinguished at this time. Solid soaps contained sodium, while liquid ones contained potassium. Potassium was extracted from some on-shore plants in Gaul [1].

Starting from the middle ages, people used soap for scientific experiments. For example, alchemist Albertus Magnus in 1260 utilised soap to obtain arsenic by recuperation [1].

The first recorded studies of soap films were carried out in the late seventeenth century by Hooke [2] and Newton [3]. They employed soap bubbles to investigate the reflection, refraction and colours of light. In doing so they were the first to document the so-called 'black' and 'multiple black' films (thin portions of soap bubble walls, which reflect very small amounts of light).

In 1741, Geoffroy proved experimentally the statement of Plinius that solid soaps consist of sodium cells, while liquid soaps consist of potassium cells.

The beginning of the nineteenth century saw the start of the era of large-scale industrial soap production. In 1811, Chevreul studied the process of soap formation, and elaborated the main principles of rational soap fabrication. In 1828, Joseph d'Arcet perfected the French soap industry [1].

Broad everyday-life and industrial uses of soap stimulated further scientific investigations of soap films. In 1873, Plateau [4] made observations of soap films suspended on various supporting wire frames and was able to obtain macroscopic films of different shapes. He outlined a specific role for the thicker part of the soap film neighbouring the supporting frame. This part was then called, in his honour, the 'Plateau border'. In 1881 and 1883 Reynold and Rücker [5, 6] and later, in 1906, Johnnott [7] measured the thickness of the 'black' portions of soap films by different techniques, and obtained corresponding results.

In parallel with the experimental investigations, theoretical studies were also in progress. Thus in 1867, Dupré [8] and, in 1891, Lord Rayleigh [9] published independent theoretical studies of hole expansion during free-standing liquid film rupture. In 1886, Reynolds [10] elaborated the macroscopic theory of thinning of a liquid film with rigid boundaries.

The first studies on soap films were reflected in monographs written by Boys in 1890 [11] and Lawrence in 1929 [12].

Approximately at the same time, liquid crystals were discovered: in 1888, thermotropic mesophases, by Reinitzer and Lehmann; and in 1916, lyotropic mesophases, by Sandqvist (see, for example, [13]).

Perrin [14], who in 1918 observed the stepwise thinning of soap films prepared from concentrated soap solutions, was familiar with the discovery of liquid crystals. This allowed him to assume that the observed phenomena was due to the layered structure of the film. Perrin went even further, supposing the ordered molecular organisation inside each of these layers. He called this stepwise thinning 'stratification'. He was the first to use the optical interferential method for measurements of soap film thicknesses.

First, experimental observations of free-standing films, prepared from thermotropic liquid crystals, were carried out by Friedel in 1922 [15]. He showed that a free surface imposes the direction of the preferred orientation of the mesophase molecules.

After about 30 years break from studying freely suspended liquid films, a considerable amount of work on this subject appeared, starting from the end of the 1940s. This was partially stimulated by the growing production of soap and by the widening of the industrial and everyday-life applications of soap products. In this connection, it is sufficient to remember, for example, the creation of the first shampoo 'Dop' made in 1928 by Mme. Eugène Schueller—the founder of 'L'Oréal'—and further improvement and broad commercialisation of this product in the 1930s to 1950s (see, for example, [16]).

Thus in 1953, Derjaguin and Titievskaya [17] and in 1957, Scheludko [18] elaborated the first experimental set-ups allowing visual microscopic observations of free-standing liquid films under controlled thermodynamic conditions (pressure inside the film, film diameter, etc.). In 1959, Mysels *et al.* [19], proposed the classification of soap films based on the character of their surface hydrodynamics.

Presently, a great number of scientific laboratories in the world deal with freely suspended amorphous and liquid crystalline films. Among the numerous directions of study are the following, which are the most actively developing:

– structure of thin films and membranes;
– different mesophases and phase transitions;
– two-dimensional hydrodynamics and defects;
– stratification, film thinning, rupture and surface viscoelasticity.

1.2 LIQUID CRYSTALLINE STATE OF MATTER

1.2.1   GENERAL NOTIONS

As has already been mentioned in the Preface, from the point of view of their structure, liquid crystals or mesophases occupy an intermediate position between solid crystals and amorphous liquids. They consist of structural elements having most often a prolonged form: e.g. long organic molecules or molecular aggregates. The long (and sometimes short) axes of these aggregates can be ordered parallel to a certain direction: thus long-range orientational ordering appears. In contrast, long-range ordering in the positions of the centres of gravity of the structural elements (which is responsible for the formation of the solid crystalline lattice) can be absent, or present only in the direction of one or two co-ordinate axes. The latter structural feature allows liquid crystals to flow like ordinary fluids, i.e. to be really liquid.

More than a century after the discovery of liquid crystals (see also the previous paragraph) a great variety of mesophases, which differ from one another by the molecular structure and physical properties, has been found. All chemical compounds that can form liquid crystalline phases are called mesomorphic substances. They are divided into two groups: thermotropic and lyotropic.

The first group was primarily discovered and is already quite well studied. The substances of this group, consisting usually of prolonged shape organic molecules (of about 10–15 Å in length), form mesophases in certain temperature ranges; and phase transitions may occur when the temperature is changed. An example of a mesomorphic substance of this class is N-(p-methoxybenzylidene)-p-butylaniline (MBBA):

The compounds of the second group are investigated to a lesser extent. These are the surface active substances (or surfactants), the molecules of which consist of a hydrophilic (often, polar) head and a lipophilic hydrocarbon tail. Due to the different relation to water of the end groups of their molecules, such substances are often called 'amphiphilic'. An example is the ammonium oleate shown at the top of page 4 (overleaf).

The amphiphilic substances form mesophases, when being diluted in sufficient concentrations in water or other solvents. Such mesophases usually consist of different molecular aggregates formed by surfactant molecules. Liquid crystalline phases in this case appear in a certain concentration range, and phase transitions may occur when the concentration is altered.

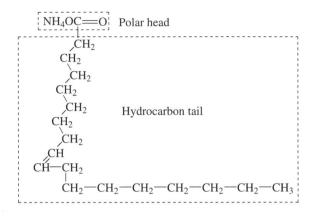

1.2.2 THERMOTROPIC MESOPHASES AND THEIR SYMMETRY

Let us first briefly describe the molecular structure and symmetry of some thermotropic liquid crystalline phases in the framework of the so-called continuum (or phenomenological) theory. Under the condition of small deformations, it considers liquid crystals as continuous, uniform and anisotropic media. Solid crystals are treated by the macroscopic crystallophysics in the same way.

This approach assumes a very important role for the point symmetry, which determines all physical properties of the medium in accordance with the famous rule of Neimann. This rule tells that 'the elements of the point symmetry group of any physical property of a substance must include the elements of the point symmetry group of the substance itself'. This means that the point symmetry group of the medium makes some restrictions on the form of tensors describing the physical properties of this medium.

The point symmetry of liquid crystals can be found by means of consideration of mutual correlations between the positions of their neighbouring molecules (or molecular aggregates), described by the so-called correlation functions. As a simple example we will consider the couple-correlation function and will find its point symmetry.

Let molecule 1 be oriented parallel to the z-axis in the rectangular co-ordinate system. The orientation of molecule 2 is described by the unit vector a parallel to its long axes and having the co-ordinates ($\sin \theta \cos \varphi$, $\sin \theta \sin \varphi$, $\cos \theta$), Fig. 1.1(a). The correlation function $f(\theta, \varphi)$ determines the mutual orientation of molecules 1 and 2. If the orientation of the molecule 1 is fixed, the value $f(\theta, \varphi)dW$ is the probability that the molecule 2 is situated inside a small angle $dW = \sin\theta\,d\theta\,d\varphi$ near the direction (θ, φ).

Suppose that the correlation function $f(\theta, \varphi)$ does not depend upon the angle φ. Moreover, we will consider molecules 1 and 2 to be non-polar, i.e. $f(\theta) = f(\pi - \theta)$. An example of such a correlation function is schematically represented in Fig. 1.1(b). It is clear from this figure that the most probable orientation of molecule 2 with respect to molecule 1 will be with their long axes parallel. This corresponds to the structure of the uniaxial nematic liquid crystal (NLC) consisting of non-polar molecules. The point symmetry of this mesophase (and of its correlation function) is ∞/mm [Fig. 1.2(a)]. It is evident that in nematics only the long-range orientational ordering of molecules (or director) exists; the long-range positional ordering is absent in all directions.

Macroscopically, the orientation of molecules of liquid crystals and, particularly, of the NLC is characterised by the director $\boldsymbol{n}$, which is the unit vector showing the average orientation of the long molecular axes in some small macroscopic volume of the mesophase: $\boldsymbol{n} = \langle \boldsymbol{a}_i \rangle$ (i denotes the number of molecules).

The just-described uniaxial non-polar nematic is the mesophase with the lowest ordering. The molecules of this liquid crystal can be modelled by the rotational ellipsoids or rods. The liquid crystalline molecules, however, can exhibit more complex geometric forms (for instance, the ellipsoid with three unequal main axes), or be polar, chiral, etc. In all these cases more complex correlation functions should be considered. Nevertheless, it is quite easy to obtain the point symmetry groups of practically all liquid crystals, considering only the forms of mesophase molecules and their mutual orientation, without the analysis of correlation functions. In most real liquid crystals the molecules are packed to ensure contact with the maximum

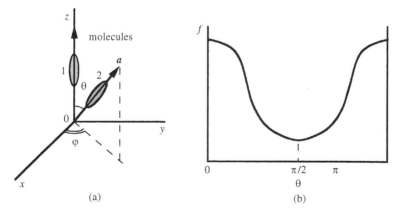

Fig. 1.1 (a) To the notion of the correlation function. (b) Correlation function with point symmetry ∞/mm

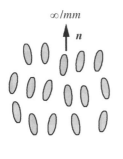

(a)

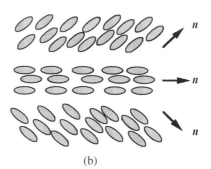

(b)

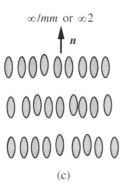

(c)

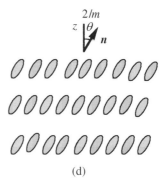

(d)

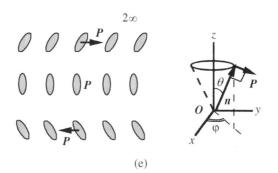

(e)

Fig. 1.2 Structure of low-ordered thermotropic liquid crystalline phases: (a) nematic; (b) cholesteric; (c) smectic A; (d) smectic C; (e) smectic C*

possible number of neighbouring molecules. This is the so-called 'statistical most dense packing' (see, for example, [20–22]). This 'dense packing' is not very different from the disordered orientation of the transverse cross sections of the mesophase molecules, i.e. the molecules of liquid crystals are not restricted in their rotations around their long axes. The point symmetry groups of different mesophase molecules (having prolonged forms) are the following: N, $N2$, N, N/m, Nm, N_i, N/mm, where N are simple symmetry axes of order N, N_i are the inversion axis of the order N. Taking into account the just-mentioned free rotation, it is found that the symmetry of most mesophases is described by the five limit point symmetry groups: ∞, $\infty\,m$, ∞/m, $\infty 2$, ∞/mm.

In general, the symmetry of liquid crystals may be described by the five limit groups mentioned above and 27 crystallographic point groups, which are the subgroups of these limit groups. These crystallographic point groups will describe the non-densely packed mesophases. The existence of the liquid crystals described by non-crystallographic groups (with symmetry axes of the fifth, seventh and higher order) is also, in principle, possible. However, experimentally, such mesophases have not yet been observed.

The second mesophase we would like to mention is the cholesteric liquid crystal. This is a nematic consisting of so-called chiral molecules: molecules without a centre of symmetry and planes perpendicular to the symmetry axes. Such molecules always form a spiral spatial structure. The pitch of the cholesteric spiral is usually of the order of the wavelength of visible light, which is why they often demonstrate a selective colourful light reflection. The point groups describing cholesterics are $\infty 2$ or ∞ [Fig. 1.2(b)].

The most ordered mesophases are the so-called smectics. The molecules of these liquid crystals form equidistant parallel molecular monolayers (or bilayers), i.e. the positional molecular ordering is in the direction perpendicular to these layers. Smectic liquid crystals are quite variable. The smectic A compounds have the easiest molecular organisation. In these liquid crystal the long molecular axes are perpendicular to the planes of the layers, and the centres of mass of molecules are distributed chaotically in each layer plane. As a consequence, the smectic A layers have a two-dimensional nematic structure. The smectics A with a symmetry centre will be described by the ∞/mm point group, while the smectics A without a symmetry centre are described by the $\infty 2$ point group [Fig. 1.2(c)].

The smectic C liquid crystal exhibits the same molecular structure as smectic A phase with the difference that all its molecules are tilted at some angle with respect to the layer normal. The presence of this tilt makes smectic C phase optically biaxial. This mesophase is described by the $2/m$ point symmetry group [Fig. 1.2(d)].

The smectic C* phase has the same molecular organisation inside each layer as smectics C. This mesophase consists of chiral molecules, often bearing electrical dipoles. The resulting dipole moment P of each molecular layer is perpendicular to the layer normal. Due to the chirality of molecules, the smectic C* mesophase exhibits a spiral structure: the polarisation P rotates as one moves from one layer to another in the direction of the layer normal. As a result, the whole bulk of this mesophase is electrically neutral. The point symmetry group of the smectic C* liquid crystal is $\infty 2$ [Fig. 1.2(e)]. Note that one can polarise a smectic C* by applying the electric field in some direction perpendicular to the layer normal, i.e. to obtain the spontaneous electric polarisation, $\mathbf{P}_{SP}$. This field will unscrew the spiral structure and will change the symmetry of this mesophase to 2 (see also Section 1.4.4).

Many smectics exhibit a two-dimensional positional ordering inside each molecular layer. These mesophases are called smectics with structured layers. The whole variety of such smectics is not yet completely studied, and we will restrict ourselves to only a short description of some mesophases, for which the data on free-standing films exist.

In the smectic B phase, the long molecular axes are perpendicular to the smectic planes, and the mass centres of molecules form a hexagonal lattice in each plane. The point symmetry group of this mesophase is $6/mmm$ [Fig. 1.3(a)].

The in-plane structures of the smectic F and the smectic I phases are similar to those of the smectic B phase with the difference that their molecules are shifted in the directions shown in Figs. 1.3 (b,c) by the arrows. As a result, the long molecular axes remain perpendicular to the smectic layers, and pseudo-hexagonal in-plane ordering appears. Both these mesophases have a monoclinic symmetry (m point symmetry group).

Note, finally, that the above-described classification of thermotropic mesophases is very approximate. It does not possess the main property of any precise scientific classification: *the possibility of prediction*. Indeed, it is not evident from this classification which new types of liquid crystals may be found. Moreover, a lot of novel mesophases are being discovered experimentally; and sometimes it is very difficult to refer the new liquid crystals to some of the already described classes.

More information on the structure of thermotropic mesophases and their physical properties can be found, for instance, in References [20–30].

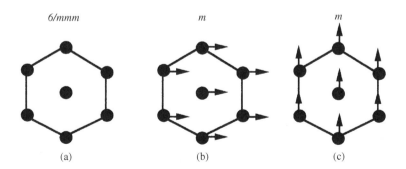

Fig. 1.3 Molecular organisation of some smectics with structured layers (adapted from [29]): (a) smectic B; (b) smectic F; (c) smectic I. Arrows indicate the direction of shift of the in-plane hexagonal lattice

1.2.3 LYOTROPIC MESOPHASES AND THEIR SYMMETRY

Lyotropic mesophases are quite numerous and their classification is even more imperfect than that of thermotropic liquid crystals. This is why we will restrict ourselves (as was done in the previous paragraph for thermotropic smectics) to a description of those lyotropic mesophases for which some experimental data on freely standing films already exist.

At low concentration c of a surfactant in some solvent (water, for example), the surfactant molecules form a monolayer at the air–liquid interface, Fig. 1.4(a). As c increases, the surfactant molecules start to penetrate the bulk of the solution, and at some value of c, which is called the critical micelle concentration (CMC), the micelles are formed in the bulk. The CMC is usually of the order of 10^{-2} mol/l. These micelles can be of different forms: circular, cylindrical, elliptical, etc. In Fig. 1.4(b), for instance, the structures of the circular form of micelle and of the isotropic solution, consisting of such micelles, are schematically represented.

When c reaches the values of the order of 1 mol/l, the proper lyotropic liquid crystalline phases may be formed. The quite densely packed circular-form micelles can arrange themselves, for example, in a centred cubical lattice. This mesophase is optically isotropic ($m3m$ symmetry) and is called sometimes the smectic D liquid crystal, Fig. 1.4(c).

The surfactant molecules may also be organised in the bulk of a solution in parallel, equidistant, molecular bilayers (lamellas) with a diminishing distance l between them with the growth of c. The symmetry of this lamellar mesophase is obviously the same as the centre-symmetrical smectics A, ∞/mm, Fig. 1.4(d).

For $c \approx 1$ mol/l, l is usually quite large, of the order of several thousand Å. This is why this liquid crystal is called the swollen lamellar phase.

More details on lyotropic liquid crystals may be found, for example, in References [31–36].

1.2.4 DEFECTS IN MESOPHASES

We have just considered the schematic, ideal structure of the main thermotropic and lyotropic mesophases. However, the real structures of liquid crystals are often far from ideal; a number of different topological defects are usually present. All defects observed in mesophases may be divided into three groups: point, linear and wall. The point defects are very common in solid crystals. These are, firstly, the molecular-size defects of Shotky and Frenkel; secondly, the defects connected with the local rotation of separate molecules. Analogous defects may also occur in mesophases. It is evident that in both solid and liquid crystals the sizes of the point defects are very small, so they cannot be detected by optical methods. However, point defects observed by means of an optical microscope sometimes occur in liquid crystals. They are connected with strong distortions in the director field. An example is the point defects at a free surface of a nematic liquid crystal shown in Fig. 1.5(a).

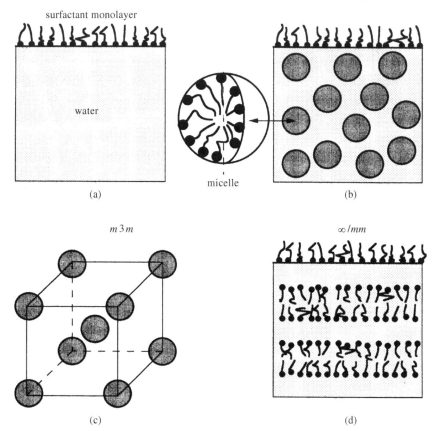

Fig. 1.4 Some lyotropic phases and mesophases: (a) monolayer of the surfactant molecules at the water–air interface; (b) circular-form micellar phase; (c) circular-form micellar cubic mesophase (smectic D); (d) lamellar mesophase

Linear defects are characterised by the appearance of a singular line. In mesophases one can observe either linear defects characteristic of solid crystals (dislocations, i.e. breaks in the continuity of the positional ordering); or linear defects, which rarely occur in solid crystals (disclinations, i.e. breaks in the discontinuity of the orientational ordering).

Dislocations often occur in liquid crystals having a layered structure (thermotropic smectics, lamellar phases). An example of a so-called edge dislocation in a smectic A liquid crystal is represented in Fig. 1.5(b).

Disclinations—the most typical defects in nematic liquid crystals—are extremely variable in their morphology. An example of the so-called wedge disclination is shown in Fig. 1.5(c).

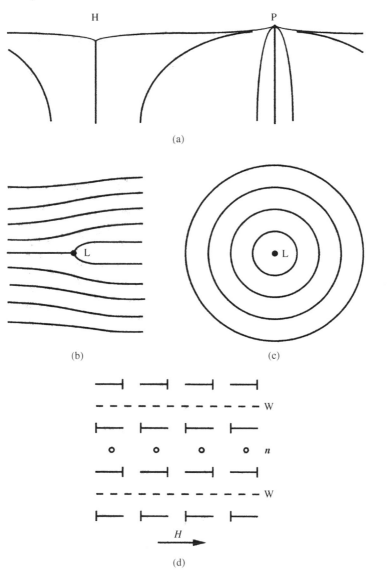

Fig. 1.5 Examples of defects in liquid crystals (adapted from [20–22, 38] by permission of Oxford University Press and of Gordon and Breach, Science Publishers): (a) point defects at the free surface of a nematic (H = hole, P = pick); (b) edge dislocation in a smectic A (L = the dislocation line perpendicular to the Figure plane); (c) wedge disclination in a nematic (L = the disclination line perpendicular to the Figure plane); (d) the 180° Bloch walls (W) in a nematic in a magnetic field H. For (a), (c), the nematic director represents a tangent at every point of the solid lines; for (b), the smectic layers are shown by solid lines; for (d), *n* is the director

Finally, the wall defects in mesophases are characterised by a singular surface. Typical examples of such defects in mematics are the 180° walls, which appear under the action of external magnetic field, or the walls between the two inversely tilted director orientations, formed in the course of the Fredericks transition (see also below in this chapter and Chapter 6). The 180° Bloch wall is represented in Fig. 1.5(d).

The interested reader can find more detailed information on defects in mesophases in, for instance, References [20–22, 37, 38].

1.3 THEORY OF ELASTICITY OF LIQUID CRYSTALS

Here we will consider the continuum theory of elasticity of ideal mesophases, which do not contain defects.

1.3.1 ORDER PARAMETERS

The compression and decompression deformations of a mesophase are, usually, very small, while shear deformations lead to a flow. Thus the main types of deformations realised in liquid crystals, are connected with the deviations in the orientation of the director.

To characterise these deformations, let us firstly describe quantitatively the long-range orientational ordering of mesophases. Consider a small macroscopic volume of a liquid crystal, containing N molecules. Their ordering can be described by the local averages from the quadric combinations, formed by the projections of the vectors a on the axes of the local co-ordinate system, connected with the considered volume of the mesophase:

$$Q_{ij} = \frac{1}{N}\sum_{N}(a_i a_j - \tfrac{1}{3}\delta_{ij}). \qquad (1.1)$$

Here the polar symmetric tensor of the second rank, Q_{ij}, is called the tensor order parameter, and δ_{ij} is the Kronecker symbol.

Indeed, if the molecules of a mesophase are situated chaotically, then $Q_{ij} = 0$; with an increase in ordering Q_{ij} also grows and reaches a value of 1 for the complete ordering. Q_{ij} can also be expressed by the components of the director n:

$$Q_{ij} = Q(n_i n_j - \tfrac{1}{3}\delta_{ij}). \qquad (1.2)$$

In the formula (1.2), Q is the scalar number, which gives the portion of molecules oriented with their long axes in a given direction. Q can be written with the help of the correlation function (see earlier in this chapter).

$$Q = \int f(\theta) \tfrac{1}{2} (3 \cos^2 \theta - 1) d\Omega$$
$$= \tfrac{1}{2} \langle (3 \cos^2 \theta - 1) \rangle = \langle P_2(\cos \theta) \rangle. \tag{1.3}$$

Here $\langle P_2(\cos \theta) \rangle$ are the Legendre polynomials of the second order.

If the correlation function is the same, as shown in Fig. 1.1(b), then it is evident from Eq. (1.3) that $Q = 1$ for a complete ordering (when all the long molecular axes are mutually parallel) and $Q = 0$ for a chaotic orientation of molecules. Thus the scalar value Q characterises the molecular ordering of the nematic mesophase and, hence, is called the 'scalar order parameter'.

In our further consideration we will need to describe the molecular ordering in some other mesophases. For this the scalar order parameter Q or some other more complex order parameters may be used.

For example, cholesteric liquid crystal Q can be introduced for each conditional 'layer'.

In the smectic A phase, the molecular ordering in each layer can also be described by Q, but the additional order parameter, characterising the layered ordering, appears. Indeed, if we consider the density ρ as the number of molecules in a unit volume (moving parallel to the director $\boldsymbol{n}$), then in nematics ρ will be constant, Fig. 1.6(a). In contrast, in the smectic A phase, due to the presence of the molecular layers, this density will be a periodical function of the co-ordinate z, perpendicular to the layers, Fig. 1.6(b):

$$\rho(z) = \rho(z + 1). \tag{1.4}$$

Here l is the distance between the smectic layers.

Let $z = 0$ be the co-ordinate of the centre of a molecular smectic layer. Then the periodic function (1.4) can be decomposed into the Fourier components as follows:

$$\rho(z) = \sum_q \rho_q \cos qz, \tag{1.5}$$

where $q = 2\pi m / l$ and $m = 0, 1, 2, \ldots$.

Restricting ourselves to the first two terms in Eq. (1.5), we can write

$$\rho(z) = \rho_0 + |\Psi| \cos \left(\frac{2\pi z}{l} \right), \tag{1.6}$$

where ρ_0 is the constant density in the nematic phase, $|\Psi|$ is the amplitude of the density appearing in the smectic A phase.

The quantity $|\Psi|$ is the additional order parameter for the smectic A phase: $|\Psi| = 0$ for nematics and $|\Psi| \neq 0$ for smectics A.

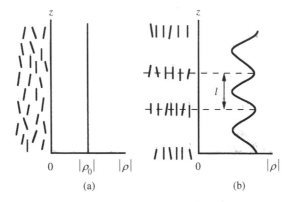

Fig. 1.6 The molecular densities for nematic (a) and smectic A (b) liquid crystals as a function of z co-ordinate. [Adapted from Reference [22]]

The molecular ordering in smectics C is described by three order parameters: Q, $|\Psi|$ and the polar tilt angle θ [Fig. 1.2(d)]. Indeed, $\theta \neq 0$ in the smectic C phase and $\theta = 0$ in the smectic A phase.

For chiral smectic C* liquid crystals the molecular ordering is also characterised by three order parameters: Q, $|\Psi|$ and Θ. Since the director in smectic C* rotates as one moves from one layer to another, the third order parameter Θ will be the complex number depending on both polar θ and azimuthal φ angles (e.g. Ref. [38]), Fig. 1.2(e):

$$\Theta = \theta \exp(i\varphi), \tag{1.7}$$

where θ is assumed to be small ($\theta \ll 1$).

In the case of the structured smectics, the situation is again more complex: an additional, so-called bond-orientational (BO) order parameter, Y, appears. It characterises the degree of ordering of physical bonds between smectic molecules inside one smectic plane. For the concrete case of the in-plane hexagonal ordering (e.g. smectic I) Y (or, more precisely, Y_6) can be written as (e.g. References [39, 40]), Fig. 1.7:

$$Y_6(r) = \exp[i6\omega(r)]. \tag{1.8}$$

Here $\omega(r)$ is the angle between the bond connecting neighbouring molecules and a reference (x) axis, and r is the in-plane radius vector.

Note, finally, that the notion of the order parameter may be applied not only to the bulk of a liquid crystalline film, but also to its surface areas. In this latter case we deal with the surface order parameter, which, as we will see later (Chapter 5), can differ from the bulk one.

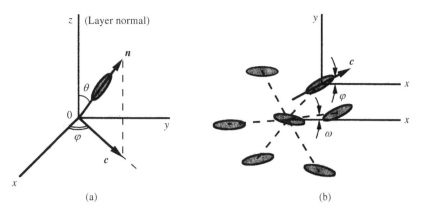

Fig. 1.7 (a) Director **n** of the smectic I liquid crystal. (b) The bond ordering in the smectic (*xy*) plane. **c** is the projection of **n** upon this plane. (Reproduced with permission from Reference [40]. Copyright 1992 American Physical Society)

1.3.2 FREE ENERGY

It is very important to know the elastic free energy (or thermodynamic potential) of a liquid crystalline film ϕ, which is usually calculated per unit area: $\Phi = \phi/S$ (S is the area of the film surface). Indeed, the minimisation of Φ with respect to the co-ordinates of the director **n** allows one to find the equilibrium distribution of **n**. Let us obtain a general expression for Φ.

Suppose that all the molecules of a liquid crystal are initially oriented parallel to some direction ($Q = 1$), and that only the deviations from the initial orientation of the director due to external actions (i.e. orientational deformations) are possible. These deformations can be expressed by the derivatives of the order parameter upon the co-ordinates (see Eq. 1.2):

$$\frac{\partial Q_{ij}}{\partial X_k} = n_i \frac{\partial n_j}{\partial n_k} + n_j \frac{\partial n_i}{\partial n_k} = a_{ijk}. \qquad (1.9)$$

Here the polar third-rank deformations tensor a_{ijk} contains 27 components, of which only 18 are independent.

Let us choose the co-ordinate system so that the orientation of the director in the non-deformed state will be parallel to the x_3 axes. Then, neglecting the small terms in a_{ijk}, which are of the order of magnitude of $\mathbf{n}^2$, we find that the orientational deformations in liquid crystals are characterised by the following components:

$$a_{3jk} = \frac{\partial Q_{3j}}{\partial x_k} = n_3 \frac{\partial n_j}{\partial x_k}, \qquad (1.10)$$

where $j = 1, 2$; $k = 1, 2, 3$ and $n_3 = 1$.

According to this formula, three different types of deformations are possible in liquid crystals: splay (the coefficients a_{311} and a_{322}), twist (a_{312} and a_{321}) and bend (a_{313} and a_{323}). Fig. 1.8.

In a general form Φ can be written as

$$\Phi = \Phi_0 + \Delta\Phi, \tag{1.11}$$

where Φ_0 is the free energy in the non-deformed state and $\Delta\Phi$ is the additional free energy connected with the deformation. In practice, it is more convenient to use the density F of the free energy:

$$\Phi = \int_v F dV, \tag{1.12}$$

where V is the volume of the sample.

Considering the orientational deformations to be small, one can decompose F to the powers of a_{3jk} (see Eq. 1.10):

$$F = F_0 + k_{3ij}a_{3ij} + \tfrac{1}{2}K_{3ij3kl}a_{3ij}a_{3kl}, \tag{1.13}$$

where F_0 is the free energy density of the non-deformed liquid crystal.

The coefficients k_{3ij} and K_{3ij3kl} in the Eq. (1.13) are, correspondingly, the components of the polar third- and sixth-rank tensors. Tensor K_{3ij3kl} is invariant with respect to the index displacements (i.e. $K_{3ij3kl} = K_{3kl3ij}$), while tensor k_{3ij} is non-invariant. The coefficients k_{3ij} and K_{3ij3kl} are called the elastic or Frank constants.

It is obvious that the symmetry of the medium should make some restrictions on the view of the elastic constant tensors. For example, in the case of a nematic liquid crystal (∞/mm symmetry), Eq. (1.13) can be rewritten as

$$F = F_0 + \tfrac{1}{2}K_{11}(\mathrm{div}\ \boldsymbol{n})^2 + \tfrac{1}{2}K_{22}(\boldsymbol{n} \cdot \mathrm{rot}\ \boldsymbol{n})^2$$
$$+ \tfrac{1}{2}K_{33}(\boldsymbol{n} \cdot \mathrm{rot}\ \boldsymbol{n})^2. \tag{1.14}$$

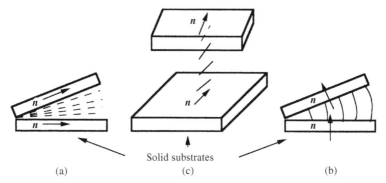

Fig. 1.8 The orientational deformations in liquid crystals (adapted from References [20, 21] by permission of Oxford University Press): (a) splay; (b) twist; (c) bend

Here the elastic constants are represented in the two-index form, as is common in the continuum theory of elasticity: $K_{11} = K_{311311}$ (splay), $K_{22} = K_{312312}$ (twist), $K_{33} = K_{313313}$ (bend).

In a real experimental situation, the ratio between the elastic constraints K_{11}, K_{22} and K_{33} for nematics is not very large, at maximum double. This is why in many cases one can restrict oneself by the so-called one-constant approximation, i.e. to assume that $K_{11} = K_{22} = K_{33} = K$. This simplifies considerably the expressions for the free energy. For instance, Eq. (1.14) will have the form

$$F = F_0 + \tfrac{1}{2}K[(\text{div } \boldsymbol{n})^2 + (\text{rot } \boldsymbol{n})^2]. \tag{1.15}$$

For the smectic A liquid crystal (∞/mm, $\infty 2$ symmetries), Eq. (1.13) will have an even easier form

$$F = F_0 + \frac{1}{2}B\left(\frac{\partial u}{\partial z}\right)^2 + \frac{1}{2}K\left(\frac{\partial^2 u}{\partial x^2} + \frac{\partial^2 u}{\partial y^2}\right)^2. \tag{1.16}$$

Here u is the displacement value along the z-axis, perpendicular to the smectic layers, B is the compressibility module and K is the splay elastic constant.

Eq. (1.16) means that the twist and bend deformations are prohibited in the smectic A mesophase. Note that in smectics C(C*), these deformations are permitted.

1.4 ELECTRO- (MAGNETO-) OPTICAL EFFECTS IN LIQUID CRYSTALS

Characteristic physical phenomena in mesophases are the change of optical properties under the action of an external electric (or magnetic) field. These phenomena, which are usually called electro- (magneto-) optical (EMO) effects (see, for example, [20–22, 25, 26]), find a great number of practical applications (for instance, in information displays, optical modulators, etc.). The EMO effects in thermotropic mesophases are the most well-studied and find the largest areas of application. Consider briefly some examples of these phenomena in nematics (Fredericks' transition, flexoelectric effect, electrohydrodynamic instabilities) and in smectics C* (linear electrooptical effect). These EMO effects will be discussed further on (Chapter 6) in the connection with FSLC films.

1.4.1 THE FREDERICKS TRANSITION

The Fredericks transition, already mentioned above in connection with the defect walls, is the change of the director orientation, induced by the interaction of an external electric (E) or magnetic (H) field with, respectively, the dielectric ($\Delta\varepsilon = \varepsilon_\parallel - \varepsilon_\perp$) or diamagnetic ($\Delta\chi = \chi_\parallel - \chi_\perp$) anisotropies of a nematic. Indices '$\parallel$' and '$\perp$' refer to ε or χ measured in the direction parallel and perpendicular to the nematic director, respectively.

The Fredericks effect exhibits a threshold character, i.e. the director field deformation starts when $E > E_F$ or $H > H_F$ (E_F and H_F are, respectively, the threshold electric and magnetic fields). A schematic director distribution in the course of this effect, induced by an electric field for a nematic with $\Delta\varepsilon > 0$ ($\Delta\chi$ is always positive for nematics), is represented in Fig. 1.9(a). In the case shown in this Figure, for the nematic director strongly fixed at the bounding solid plates, E_F is given by the following simple formula (see, for example, [20–22, 25, 26]):

$$E_F = \frac{\pi}{h}\left(\frac{4\pi K_{33}}{\Delta\varepsilon}\right)^{1/2}. \tag{1.17}$$

Here h is the thickness of the nematic layer.

Eq. (1.17) can be easily rewritten for a magnetic field by the following substitution: $(\Delta\varepsilon/4\pi)E^2 \rightarrow \Delta\chi H^2$.

1.4.2 FLEXOELECTRIC EFFECT

The flexoelectric effect is the appearance of the so-called flexo-deformation under the action of an external electric field. It can be observed in nematics consisting of polar molecules with anisotropy of form. A schematic mechanism of this phenomenon for a nematic with banana-shaped molecules and negative dielectric anisotropy is shown in Fig. 1.9(b). In the absence of an electric field ($E = 0$), all molecular dipoles are disordered in the plane perpendicular to the director n, and the average electric polarisation of a nematic sample $P = 0$. If an electric field is applied perpendicular to n, the moleculecular dipoles will be oriented parallel to E (and now $P \neq 0$ the flexoelectric polarisation appears). This obviously will be accompanied by the reorientation of the molecules and, hence, by the flexo-deformation of the director field. In the case where the director is not fixed at the boundaries and for small deformations ($\theta \ll 1$), the average birefringence over the nematic layer thickness $\langle \Delta n \rangle$ will be given by the following formula (e.g. [20–22, 25, 26], see also Eq. 3.7):

$$\langle \Delta n \rangle = \frac{n_o}{24}\left(1 - \frac{n_o^2}{n_e^2}\right)\frac{e_{33}^2}{K_{33}^2}e^2 h^2. \tag{1.18}$$

Here n_o and n_e are, respectively, the ordinary and extraordinary refractive indices of the nematic; e_{33} is its flexoelectric coefficient.

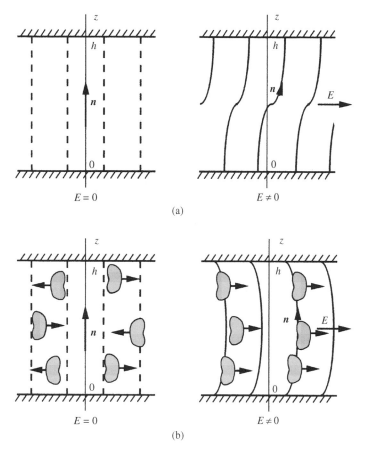

Fig. 1.9 Schematic geometries of some field-induced effects in thermotropic nematics (from Reference [38] by permission of Gordon and Breach, Science Publishers): (a) Fredericks' transition; (b) flexoelectric effect

1.4.3 ELECTROHYDRODYNAMIC INSTABILITIES

Consider, following References [20–22, 25, 26], some electrohydrodynamic (EHD) instabilities in thermotropic nematics. Most of the experimental observations of the EHD effects have been made in nematics with negative dielectric anisotropy ($\Delta\varepsilon < 0$) and positive anisotropy of the electric conductivity ($\Delta\sigma > 0$); MBBA, for example. If a layer of such a nematic (with a typical thickness of about 30 m) is placed between two semitransparent electrodes, and a d.c. or a.c. low-frequency electric field is then applied, the following effects may be observed with increasing voltage U.

For low U (as a rule, about 1 V), the nematic molecules align themselves perpendicular to the electric field E, along a certain axis. Thus the nematic single crystal is being formed.

When U reaches a certain critical value U_c (of approximately 5 V), a periodic deformation is observed. In many cases, this is a simple one-dimensional distortion—the so-called Williams domains—which may exhibit a great variety of forms. The most often occurring Williams domain texture represents, in the crossed microscope Nicol prisms, a number of mutually parallel stripes, oriented perpendicular to the initial director orientation [Fig. 1.10(a)]. The period of these domains, λ, is proportional to the nematic layer thickness: $\lambda \propto h$.

When the applied electric voltage increases further, the deformation amplitudes and the liquid flow velocities also grow. This leads to the destruction of the long-range orientational ordering and to the appearance of turbulent motion. This regime is usually called 'dynamic scattering', due to the strong diffuse light scattering shown by a nematic sample.

If the frequency ω of the applied electric field increases, U_c for the EHD instability will grow, but the texture will have the form of the Williams domains. For $\omega > \omega_c$, where $f_c = \omega_c/2\pi$ is some critical frequency (of the order of magnitude of the relaxation frequency, f_r, of the ionic electric charges in the nematic film; see Eq. 1.23 below), U_c will be increased considerably, and λ will be diminished. Note that f_c usually has a value of the order of 100 Hz. As a result, the new type of EHD texture, generally called 'chevrons', may appear [Fig. 1.10(b)].

Let us briefly describe the so-called Carr–Helfrich model proposed for the explanation of the just-mentioned EHD instabilities. Consider a nematic layer with $\Delta\varepsilon < 0$ and $\Delta\sigma > 0$, and the director initially oriented along the x axis. Let the external dc electric field be submitted along the z-axis [Fig. 1.10(c)]. The texture distorted by this field exhibits small periodic splay deformation. The elastic energy grows due to this distortion. This initiates the counterbalance force tending to restore the initial texture. On the other hand, since $\sigma_\parallel > \sigma_\perp$, the electric current component, I_t, along the x axis appears. It tends to accumulate positive charges in the area of point A, which leads to the following two main effects:

1. The electric field at point B is being changed from E to $E + \delta E$. The nematic molecules at this point try to remain perpendicular to the field. As is clear from Fig. 1.10(c), this electrostatic rotational torque tends to increase the initial distortion.

2. In the vicinity of point A, the fluid is subjected to the action of the bulk force qE. This leads to a certain flow pattern shown in Fig. 1.10(c). As a result, a considerable hydrodynamic rotation torque appears at point B. This torque tends to increase the distortion.

If α is the angular amplitude of the distortion, and k is its wave-vector along the x-axis, then the rotation torque due to the elastic deformation equals $K_{33}k^2\alpha$. On the other hand, both the electrostatic and hydrodynamic rotational torques are

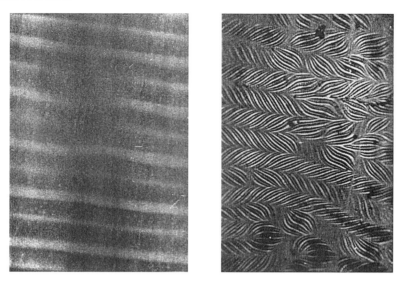

(a)   (b)

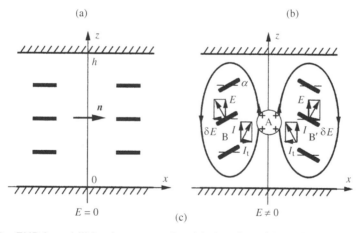

$E = 0$ $E \neq 0$

(c)

Fig. 1.10 EHD instabilities in a nematic with $\Delta\varepsilon < 0$ and $\Delta\sigma > 0$, observed by the polarising microscope (adapted from References [20, 21] by permission of Oxford University Press): (a) Williams domains; (b) chevrons; (c) Carr–Helfrich mechanism of the EHD instabilities

proportional to $E^2\alpha$. Since, experimentally, the distortion wavelength (along the x axis) is proportional to the sample thickness h, then $k \propto 1/h$.

Thus we come to the conclusion that there exists a critical electric field E_c, determined by the following equation (e.g. [20, 21]):

$$E_c^2 = \frac{\text{const.} \cdot K_{33}}{h^2}. \tag{1.19}$$

Eq. (1.19) means that the critical voltage of the formation of the Williams EHD domains, $U_c = E_c h$, does not depend on the layer thickness; the fact observed experimentally.

The precise value of the constant in Eq. (1.19) can be found by means of the quite complex two-dimensional nemato-dynamic calculations (see Chapter 6). However, its approximate value can be obtained using the one-dimensional approximation. All the parameters of the problem, such as the tilt angle α, are supposed to depend only on the transverse co-ordinate x, but not on z; the boundary conditions on both plates are not considered (only the wave-vector k is supposed to be equal to c/h, where c is a constant approximately equal to π). Then it is relatively easy to obtain the result

$$U_c^2 = \frac{U_0^2}{\zeta^2 - 1}, \tag{1.20}$$

where

$$U_0^2 = 4\pi^3 \frac{K_{33}\varepsilon_\|}{\varepsilon_\perp(\varepsilon_\perp - \varepsilon_\|)} \tag{1.21}$$

and ζ^2 is a dimensionless parameter, having the value

$$\zeta^2 = \left(1 - \frac{\sigma_\perp}{\sigma_\|}\frac{\varepsilon_\|}{\varepsilon_\perp}\right)\left(1 + \frac{\alpha_2\varepsilon_\|}{\eta_c\Delta\varepsilon}\right). \tag{1.22}$$

Here α_2 and η_c are the viscosity coefficients.

The necessary condition for the EHD instability is $\zeta^2 > 1$. For MBBA, for instance, this condition is always fulfilled: $\zeta^2 \approx 3.2$.

The same approach gives the following expression for the threshold of the low-frequency EHD instability (e.g. [20, 21]):

$$U_c(\omega) = U_0\left[\frac{1 + \omega^2\tau^2}{\zeta^2 - (1 + \omega^2\tau^2)}\right]^{1/2}. \tag{1.23}$$

Here U_0 and ζ^2 are given by the Eqs. (1.21) and (1.22), respectively; $\tau = \varepsilon/4\pi\sigma$ is the dielectric relaxation time.

It is clear from Eq. (1.23) that for $\omega \to \omega_c = 1/\tau\sqrt{(\zeta^2 - 1)}$ $U_c(\omega)$ grows considerably, which is in accordance with the experimental results.

Moreover, $1/\tau$ and, hence, ω_c are the linear functions of the electro-conductivity (see Eq. 1.23), which is also in accordance with experiment. As a consequence, for $\omega < \omega_c$, the distortion pattern corresponds to the oscillating electric charges with the static director perturbation. This is why such EHD instability is often called the 'conductivity regime'.

For the high-frequency ($\omega > \omega_c$) EHD instability, Eq. (1.22) is not valid. It was found that in this regime $E_c \propto \omega^{-1/2}$, E_c does not depend on h and $\lambda \propto \omega^{-1/2}$. This is in accordance with the experimental observations. For $\omega \gg \omega_c$, the nematic molecular structure oscillates, while the electric charges remain immovable. This is why such EHD instability is often called the 'dielectric regime'.

1.4.4 LINEAR ELECTROOPTICAL EFFECT IN FERROELECTRIC SMECTICS

The macroscopic polarisation P_{sp} of ferroelectric smectic C* liquid crystals (often called spontaneous) interacts with an external electric field, applied parally to the smectic layer planes. This (as we already know) may lead for instance, to the unwinding of the helical structure at some critical field E_{th} (the helix pitch p goes to infinity), i.e. to the change of the director orientation and thus of the optical properties of the sample [see Fig. 1.2(e)]. E_{th} is given by the following equation [25, 26]:

$$E_{th} = \frac{\pi^4 K_{22}}{4p_0^2 P_{sp}},$$ (1.24)

where K_{22} is the twist elastic constant of the smectic C*, and p_0 is its initial helix pitch (for $E = 0$).

It is clear from Eq. (1.24) that, in contrast with the above-described Fredericks transition in nematics, the electrooptical effect in smectics C* is linear with respect to the electric field E; the direction of reorientation of the liquid crystalline director depends on the sign of E.

The other important characteristic of the described phenomenon is the response time t_r (time of the complete director reorientation in the helix unwinding process). For practical applications in various information displays, one needs to make E_{th} and t_r as low as possible (see, for example, [25, 26]).

REFERENCES

1. Bloch M.A., 1940, *Chronology of the Main Events in the Domain of Chemistry*, Goskhimizdat, Moscow (in Russian).
2. Hooke R., 1757, in *History of the Royal Society of London* (ed. Brich T.), London, vol. 3, p. 29.
3. Newton I., 1704, *Opticks*, Smith & Walford, London.
4. Plateau J., 1873, *Statistique Expérimentale et Théorique des Liquides Soulis aux Seules Forces Moléculaires*, Gauthier-Villars, Paris, vol. 1, p. 163.
5. Reynold A.W., Rücker A.W., 1881, *Trans. R. Soc. London*, **172**, 447.
6. Reynold A.W., Rücker A.W., 1883, *Trans. R. Soc. London*, **174**, 645.
7. Johnnott E.S., 1906, *Philos. Mag.*, **11**, 706.
8. Dupré A., 1867, *Ann. Chim. Phys.*, **4(11)**, 194.
9. Rayleigh J.W.S., 1891, *Nature*, **44**, 249.
10. Reynolds O., 1886, *Philos. Trans. R. Soc. London*, **A177**, 157.
11. Boys C.V., 1890, *Soap Bubbles and the Forces Which Mould Them*, London Society of Promptory Christian Knowledge.

12. Lawrence A.S.C., 1929, *Soap Films*, Bell, London.
13. Sonin A.S., 1988, *Road of Century Length (From the History of Liquid Crystal Science)*, Nauka, Moscow (in Russian).
14. Perrin J., 1918, *Ann. Phys. (Paris)*, **10**, 160.
15. Friedel G., 1922, *Ann. Phys. (Paris)*, **18**, 273.
16. Rachline M., 1991, *Le saga Dop. Le romain d'un produit mythique*, Albin Michel, Paris.
17. Derjaguin B.V., Titievskaya A.S., 1953, *Kolloid. Zh.*, **15**, 316.
18. Sheludko A., 1957, *Kolloid. Zh.*, **155**, 39.
19. Mysels K.J., Shinoda K., Frankel S., 1959, *Soap Films: Studies of Their Thinning and Bibliography*, Pergamon Press, New York.
20. de Gennes P.G., 1974, *The Physics of Liquid Crystals*, Clarendon Press, Oxford, UK.
21. de Gennes P.G., Prost J., 1993, *The Physics of Liquid Crystals*, 2nd edn., Clarendon Press, Oxford, UK.
22. Sonin A.S., 1983, *Introduction to the Physics of Liquid Crystals*, Nauka, Moscow (in Russian).
23. de Jeu W.H., 1980, *Physical Properties of Liquid Crystalline Materials, Gordon and Breach, London.*
24. Vertogen G., de Jeu W.H., 1988, *Thermotropic Liquid Crystals, Fundamentals*, Springer, Berlin.
25. Blinov L.M., 1983, *Electro-Optical and Magneto-Optical Properties of Liquid Crystals*, Wiley, Chichester, UK (Russian version: 1978, Nauka, Moscow).
26. Blinov L.M., Chigrinov V.G., 1994, *Electrooptic Effects in Liquid Crystal Materials*, Springer, New York.
27. Pikin S.A., 1991, *Structural Transformations in Liquid Crystals*, Gordon and Breach,New York (Russian version: 1981, Nauka, Moscow).
28. Chandrasekhar S., 1992, *Liquid Crystals*, 2nd edn. Cambridge University Press, Cambridge, UK.
29. Gray G.W., Goodby J.W., 1984, *Smectic Liquid Crystals—Textures and Structures*, Leonard Hill, Glasgow.
30. Pershan P.S., 1988, *Structure of Liquid Crystal Phases*, World Scientific, Singapore.
31. Ekwall P., 1975, in *Advances in Liquid Crystals* (ed. G.H. Brown) Academic Press, New York, vol. 1, p. 1.
32. Skoulios A., 1978, *Ann. Phys.*, **3**, 421.
33. Charvolin J., 1984, in *Colloides et Interfaces* (eds. A.-M. Cazabat, M. Veyssie), Les Editions de Physique, Les Ulis, France, p. 33.
34. Sonin A.S., 1987, *Usp. Fiz. Nauk*, **153**, 273.
35. Vasilevskaya A.S., Generalova E.V., Sonin A.S., 1989, *Usp. Khim.*, **58**, 1575.
36. Sonin A.S., 1991, *Zh. Strukt. Khim.*, **32**, 137.
37. Kléman M., 1975, in *Advances in Liquid Crystals* (ed. G.H. Brown) Academic Press, New York, vol. 1, p. 267.
38. Sonin A.A., 1995, *The Surface Physics of Liquid Crystals*, OPA-Gordon and Breach, Amsterdam.
39. Cheng M., Ho J.T., Hui S.W., Pindak R., 1988, *Phys. Rev. Lett.*, **61**, 550.
40. Sprunt S., Spector M.S., Litster J.D., 1992, *Phys. Rev. A*, **45**, 7355.

2

Free-Standing Films—Basic Physical Notions and Properties

In this chapter we will formulate some principal definitions and briefly review the main macroscopic physical properties of both liquid crystalline and isotropic free-standing films. The information given here will be essential for further description of the structure and physical effects of freely suspended liquid crystalline (FSLC) films.

2.1 SURFACE AND BULK AREAS OF A FILM AND CORRESPONDING ORDER PARAMETERS

Consider a freely suspended liquid film, schematically represented in Fig. 2.1. This film is of disc form, having thickness h and radius R and consisting of adequate molecules or molecular aggregates with a maximal size a. Due to the action of the surface forces at the film interfaces, the molecules (or molecular aggregates) forming the film generally exhibit higher spatial and orientational ordering near the film surfaces than in its bulk. The degree of this ordering can be characterised by the already-mentioned order parameter (see Chapter 1), the value of which, obviously, will be higher in the vicinity of the film interfaces than in its bulk.

The easiest example of a FSLC film is a thermotropic nematic free-standing film. The molecular ordering in this film is characterised by the nematic order parameter Q (see formula 1.2). According to theoretical estimations (see, for example, [1, 2]), the value of Q near the surface of such a film is about 0.4–0.8, while in the bulk of the mesophase $Q = 0.3$–0.5. The main decrease of Q takes place in the surface areas at a depth h_s of about several hundred Å. These results have also been proved experimentally for free surfaces of some thermotropic nematics (e.g. [1, 3]).

A schematic co-ordinate dependence of the order parameter, $Q(z)$, for a thermotropic, nematic, free-standing film is shown in Fig. 2.2. The interfacial film parts of thickness h_s with a higher molecular ordering are usually called the 'surface areas' (see Fig. 2.1), while the value of the order parameter in these areas averaged over the

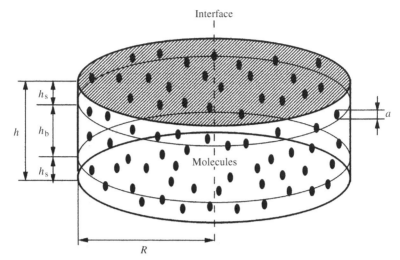

Fig. 2.1 Schematic representation of a free-standing film: h is the film thickness; h_s and h_b are, respectively, the thickness of the surface and bulk areas of the film; R is the film radius; and a is the maximal size of the film-constituting molecules (or molecular aggregates)

co-ordinate z, which is measured experimentally, is called the 'surface order parameter', Q_s:

$$Q_s = \frac{1}{h_s} \int_0^{h_s} Q(z)\mathrm{d}z. \tag{2.1}$$

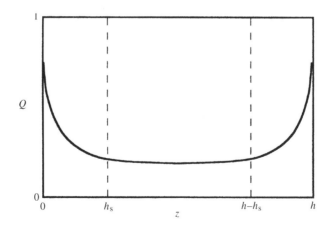

Fig. 2.2 Schematic z-co-ordinate dependence of the order parameter, Q, for a nematic free-standing film

A central portion of the film, having the thickness $h_b = h - 2h_s$, is called the 'bulk area'. The molecular ordering in this part of the film is obviously described by the 'bulk order parameter', Q_b, a value, measured experimentally, and also obtained by averaging over the co-ordinate:

$$Q_b = \frac{1}{h - 2h_s} \int_{h_s}^{h-h_s} Q(z)\mathrm{d}z. \qquad (2.2)$$

It is evident (see Fig. 2.2) that the influence of the interfaces in the surface areas of the film is pronounced, while it is negligible in its bulk area.

Note that the mean surface and bulk order parameters for other mesophases (smectic A, smectic C, etc.) may be introduced by substitution of the correspondent order parameter co-ordinate functions, $f(z)$, into Eqs. (2.1) and (2.2).

In the general case, the thickness h_s of the surface areas of free-standing films varies from several times a (10–500 Å) for amorphous films up to several thousand times a (several µm) for highly ordered liquid crystalline films [1–5].

2.2 'THICK' AND 'THIN' FILMS

The notion of the thickness of free-standing films is important for the characterisation of their physical properties. Let us determine what we mean when talking about thick and thin films.

It seems to be natural to define a thick film as a film for which $h \approx R$ and a thin one as a film for which $h \ll R$ (see Fig. 2.1). However, this definition is not quite satisfactory. Indeed, experimentally, the condition $h \ll R$ is fulfilled almost for all free-standing films. Moreover, the present definition does not reflect the molecular structure of the film and, hence, its physical properties.

A more correct definition can be given using the notion of the thicknesses of the surface (h_s) and bulk (h_b) areas of the film. Indeed, a thick film can be determined as a film with $h \gg h_s$ and a thin one as a film with $h \approx h_s$. This definition shows that a thick film consists mostly of the bulk area, while a thin one mostly of the surface areas, Fig. 2.3. Thus due to the different degree of the molecular ordering, the physical properties of thick and thin films should be different. Indeed, in many cases, thin FSLC films may be considered as two-dimensional systems, i.e. objects exhibiting specific two-dimensional physical properties.

According to our definition, the thickness of thick FSLC films varies, in a real experimental situation, between several tens of µm and several thousands of Å, whereas for thin FSLC films h lies between several hundreds and several tens of Å (see, for example, [1, 6]).

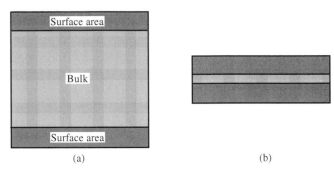

Fig. 2.3   Freely suspended films: (a) thick; (b) thin

2.3   SOME PRINCIPLES FOR THE MOLECULAR ORGANISATION OF FREE-STANDING FILMS

Thermotropic FSLC films can be formed from different thermotropic mesophases (see Chapter 1). Thin thermotropic FSLC films ($h < 1000\,\text{Å}$) can even be obtained from an isotropic (i.e. completely disordered in the bulk) phase of different mesogenes. In this latter case the surface forces may stabilise some lower-symmetrical, for instance, nematic or smectic A phases in the boundary regions of the film (for example, Reference [1]; see also Chapters 4 and 7).

Now some comments on lyotropic FSLC films. If we try to form a freely suspended film, say, from water, we will not succeed: the film will rupture immediately. The same will occur with the films prepared from many amorphous (not very viscous) fluids. However, it is evident that a film of any liquid can easily be formed between plane-parallel, flat solid walls. In this case the film is stabilised by rigid surfaces.

The surface surfactant monolayers already mentioned in the previous chapter can also play a role in such stabilising 'walls'. Indeed, as we will see later (see below and Chapter 4), such monolayers can be really rigid, or possess a considerable surface viscoelasticity, due to their ordered structure. This protects the fluid film formed between two surfactant monolayers from immediate collapse.

The free-standing films stabilised by surfactant monolayers are familiar to anyone. These are soap films. They can be prepared from solutions of surfactants in water or some other solvents. A schematic structure of a dilute soap film ($c < \text{CMC}$) is represented in Fig. 2.4(a). This film consists of a bulk water core bounded by two 'walls' of surfactant monolayers. If we increase a little the concentration of the surfactant to satisfy the condition $c > \text{CMC}$, the film consisting, for instance, of disordered circular-shaped micelles in the bulk and surfactant monolayers at the surfaces can be obtained [Fig. 2.4(b)]. These are some examples of amorphous lyotropic free-standing films.

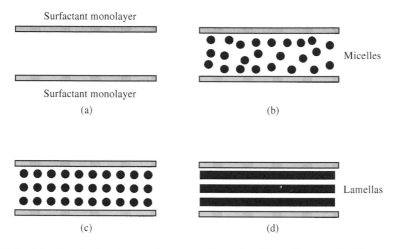

Fig. 2.4 Idealised structure of some free-standing films: (a) dilute soap; (b) disordered circular-form micelles; (c) organised circular-form micelles; (d) lamellas

Lyotropic FSLC films can be formed from highly concentrated ($c \gg$ CMC) surfactant solutions. Two most typical structures of such lyotropic liquid crystalline films are: the ordered micellar (with a cubic lattice) and, as already mentioned in Chapter 1, the lamellar, both shown in Figs. 2.4(c) and (d).

Analogous mesophase films can also be obtained by decreasing the thickness h of the just-mentioned amorphous lyotropic films with $c >$ CMC (consisting of the initially disordered molecular aggregates: micelles for example), to values of several hundreds of Å (see also Chapters 4 and 7). In this case (as for thin thermotropic FSLC films), due to the action of the surface forces, the molecular ordering increases, and some mesophase with lower symmetry (than that of an isotropic phase) can be formed. For example, micelles may organise themselves either in a cubic crystalline lattice (ordered micellar phase) [7, 8], or surfactant molecules will form parallel molecular bilayers (lamellar phase) [9], Figs. 2.4(c) and (d).

Finally, FSLC films of very low thicknesses ($h <$ several hundred Å) can, in principle, be formed even from amorphous (non-mesogenic) fluids: some organic and non-organic substances, emulsions and suspensions (e.g. [6]). Again, the molecular ordering in this case is due to the action of the surface forces.

2.4 SURFACE TENSION

As is well known, the surface tension of a liquid, γ, represents the work that is necessary to create a unit area of an interface between liquid and air (e.g. [10]):

$$\gamma = \frac{dA}{dS}. \tag{2.3}$$

Here A is the work and S is the area of the liquid–air interface.

It obvious that γ is the surface free energy density F_s of a liquid.

The presence of non-zero surface tension is the necessary condition for the existence of thermotropic FSLC films. However, lyotropic films can sometimes be stable even if $\gamma = 0$. This is, for example, the case of the swollen lamellar phase, where the bilayers are submitted to huge fluctuations governed by the curvature rigidity (e.g. [11–13]).

2.4.1 SURFACE TENSION AND ANCHORING ENERGY OF THERMOTROPIC MESOPHASES AND FREE-STANDING LIQUID CRYSTALLINE FILMS

In thermotropic liquid crystals the surface free energy density, F_s, is the sum of two parts: isotropic, $F_{si} = \gamma$, and anisotropic, F_{sa} (which depends on the director orientation at the interface) (e.g. [1, 14]):

$$F_{sa} = \frac{W}{2} \sin^2 \alpha. \tag{2.4}$$

Here α is the angle of the deviation of the director from the direction of the preferable orientation (or the 'easy orientation' axes) at the surface; W is the so-called anchoring energy: this is the energy necessary for the complete ($180°$) deviation of the director from the easy orientation axes.

W is an important physical parameter, which strongly influences many phenomena in liquid crystals, such as the EMO (electro- (magneto-) optical) effects (see [1] and Chapter 1).

Eq. (2.4) represents the anisotropic part of the surface free energy density in the so-called Rapini approximation [14]. This is the most simplified and commonly used form of the analytical expression for $F_{sa}(\alpha)$. Further on, we will utilise $F_{sa}(\alpha)$ in the Rapini form.

In a general case, the orientation of the director with respect to the easy orientation axes at the surface may be characterised by two angles: polar (θ) and azimuthal (φ) [analogously to Fig. 1.1(a)]. So the two components of the anchoring energy: polar (W_p) and azimuthal (W_a) can be distinguished.

Note that if the director is perpendicular to the surface ($\theta = 0$), its orientation is called homeotropic; if it is inclined with respect to the surface ($0 < \theta < \pi/2$), it is called tilted; and, finally, if it lies in the plane of the surface ($\theta = \pi/2$), it is called planar.

Typical experimental values of W (for nematics), contacting with solid substrates, or having a free interface, lie in the limit $W = 10^{-3} - 1\ \mathrm{erg/cm^2}$ (e.g. [1, 15]). These values are intriguingly small. Indeed, for a liquid crystal, as for any condensed medium, there is a natural estimate of the surface energy, given by (e.g. [1])

$$F_s \approx \rho v^2 a, \tag{2.5}$$

where ρ is the density, v is the velocity of sound and a is the molecular size.

Actually, $\rho v^2 a$ is the only quantity with the dimension of $\mathrm{erg/cm^2}$. Setting $\rho \approx 1\ /\mathrm{cm^3}$, $v \approx 10^5\ \mathrm{cm/s}$ and $a \approx 10^{-8}\text{–}10^{-7}\ \mathrm{cm}$, we obtain $F_s \approx 10^2\ \mathrm{erg/cm^2}$. The experimentally measured isotropic part of the surface tension in liquid crystals (and, incidentally, in any other organic liquids) is in fact of this order of magnitude. The anchoring energy determines the anisotropic part of the surface tension. All the bulk properties of liquid crystals have an anisotropic part, which is of the order of 0.1 (or even larger) of the isotropic part. Thus one should expect the values of W to be of the order of $10\ \mathrm{erg/cm^2}$, i.e. several orders of magnitude lower than that obtained experimentally. This is most likely to indicate that there exists a special surface mesophase layer (probably with a large number of defects). To solve this problem further structural investigations of the liquid crystal surfaces are needed.

It is often quite convenient to characterise anchoring by a macroscopic parameter b, having the dimensions of length and called the 'extrapolation length':

$$b = K_{ii}/W. \tag{2.6}$$

Here K_{ii} is one of the elastic constants of a nematic ($i = 1, 2, 3$).

Physically, b reflects the relation between the bulk elastic and the surface anchoring torques.

W strongly depends on temperature. For instance, for a nematic mesophase it decreases with growing T, approaching at the nematic–isotropic phase transition temperature (T_{NI}) very small, but non-zero values (~ 0.01 of its initial value in the nematic phase), and decreasing, finally, to zero at $T - T_{NI} \approx$ several K (Fig. 2.5). The non-zero values of W for $T \geqslant T_{NI}$ are due to the presence of the nematic phase in the interfacial areas of the isotropic liquid (e.g. [1, 15]).

The surface tension, γ_f, of some thermotropic smectic flat films and bubbles has been measured by techniques specially elaborated for these purposes [13, 16–22] (see also Chapter 3). It is found that γ_f has the order of magnitude of 10–50 dyne/cm for smectic A and B films [13, 16–22]. For some flat smectic A films, it is shown that γ_f does not depend on the film thickness in the following interval of the smectic layers number, 2–120 layers, and on the temperature within the interval of the existence of the smectic A phase [16–18].

The anchoring energy, W_f, has not yet been measured, however, for thermotropic FSLC films. Nevertheless, for such films, with quite large thicknesses ($h >$ several hundred Å), one should expect similar values of W_f as for the free interfaces of the corresponding bulk mesophases.

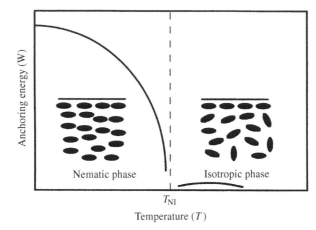

Fig. 2.5 Schematic temperature dependence of the anchoring energy, *W*, for a nematic liquid crystal (reproduced from Reference [15] by permission of Taylor & Francis)

2.4.2 SURFACE TENSION OF LYOTROPIC MESOPHASES AND FREE-STANDING LIQUID CRYSTALLINE FILMS

In the case of lyotropic liquid crystals, γ represents the free energy density of the surfactant monolayer adsorbed at the surface. The value of γ depends both on the surfactant concentration c and on the time t (e.g. [23, 24]). With growing c ($c <$ CMC) γ decreases, reaching its minimal value at $c =$ CMC, and for $c \geqslant$ CMC, γ practically does not change (Fig. 2.6). The decrease of γ is due to the growing density n of the surfactant molecules adsorbed at the interface. At $c =$ CMC, n is maximal, which corresponds to the minimal value of the surface tension. For further growth of c, the micelles are formed in the bulk, but the surface surfactant monolayer remains unchanged, i.e. γ also does not change.

The value of γ also diminishes with growing t, approaching its equilibrium value. This is caused by the gradual increase of n with time, due to the surfactant molecule diffusion from the bulk to the interface. This effect is important only for small surfactant concentrations: $c < 0.1$ CMC [23], i.e. it can often be neglected in the consideration of mesophase films.

Note, finally, that the notion of the anchoring energy, *W*, is also recently introduced for lyotropic liquid crystals, namely, for nematics consisting of cylindrical micelles [25]. For these substances the director describes the mean orientation of the long axes of micelles. This means that the surface free energy, F_s, of lyotropic nematics consists of an isotropic part (γ) and of an anisotropic part (proportional to W). The latter can, in principle, be determined analogously to that for thermotropic mesophases.

The experimental values of *W* for lyotropic nematics, however, are found to be one to two orders of magnitude lower than for thermotropic ones [25]. This discrepancy

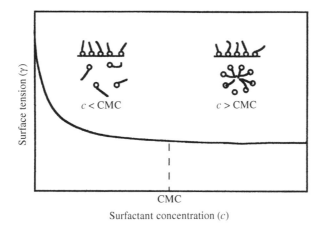

Fig. 2.6 Schematic surfactant concentration dependence of the surface tension, γ, for the surfactant solution–air interface

is probably connected with the more complex character of the interfacial interactions for micelles than for thermotropic molecules. It seems that the macroscopic approach is not sufficient to determine quantitatively the surface energy of lyotropic liquid crystals, and that the microscopic theory of anchoring is needed in this case.

We do not know the direct experimental measurements of the surface tension, γ_f, and the anchoring energy, W_f, of freely suspended lyotropic films. Nevertheless, as in the case of thermotropic FSLC films, for lyotropic films with quite large thicknesses ($h >$ several hundred Å), one should expect similar values of γ_f and W_f, as for the free interfaces of the corresponding bulk mesophases.

2.5 SURFACE VISCOELASTICITY OF SURFACTANT MONOLAYERS

Consider the peculiar viscoelastic properties of lyotropic free-standing films. Monolayers of a surface active substance, surrounding such a film, possess a special (dilational) form of elasticity such that any applied stress which tends toward local thinning or stretching of the film is rapidly opposed and counterbalanced by restoring forces generated during the initial displacement of the film material. These restoring forces increase with the amount of displacement of the film, as in the case of stretching a rubber band.

The dilational surface elasticity, as will be shown later (see Chapter 4), greatly influences the lyotropic free-standing films rate of thinning and stability.

As a local spot in the film stretches and the area of the film in that region increases (Fig. 2.7), its surface tension also increases and a gradient of tension is set up. This

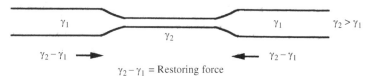

Fig. 2.7   The stretched portion of a lyotropic film: an illustration of the mechanism of film elasticity (adapted from Reference [23] by permission of John Wiley & Sons ©1989)

gradient causes liquid to flow toward the thinned spot from the thicker portion around it. This prevents the thinner spot of the film from thinning further.

Gibbs [26] proposed a quantitative description of the just-mentioned phenomenon in the terms of the surface dilational elasticity, ε. He determined ε as the local increase in surface tension, γ, for a unit of relative increase in the film area, S:

$$\varepsilon = \frac{d\gamma}{d \ln S} \tag{2.7}$$

In the most general case, the viscous and elastic properties (viscoelasticity) of a liquid film are characterised by four components: share and dilational viscous moduli and elastic modulus. Share viscoelasticity shows the resistance to the displacement of the substance at the surface, i.e. to the surface flow. Dilational, as it is clear from the above-said, shows the resistance to the decompression or compression of the substance during film thinning. This second type of film viscoelasticity was first observed by Rayleigh [27] and Marangoni [28]. The latter realised that Plateau's [29] experiments on the retarded movement of a needle suspended in the surface of a liquid did not primarily measure the surface resistance against share. He noticed that the movement of the needle caused surface contractions in front of it and dilation in its wake, and that the resulting difference in surface tension normally retarded the needle's movement rather more than any shear surface viscosity.

The latter direct measurements on the surfactant monolayers have shown that, in this case, the values of the dilational viscoelastic modulus are several orders of magnitude higher than that of the shear moduli for the same surfactant (see, for example, [30]). So further on we will neglect the share viscoelasticity and will consider only the dilational elastic properties of surface monolayers.

Neglecting the share elastic modulus reduces the surface stress tensor to the scalar quantity $\Delta\gamma$, equal to the difference between the surface tension of an interface after and before deformation.

Consider the case of small (< 1%) periodic surface deformations. Experimentally, such deformations can be obtained, for example, by generating surface waves [31, 32], by regular movements of some barrier along the fluid interface. Under the action of these waves the surface tension oscillates around its equilibrium value. The amplitude of these oscillations can be made to be small, as required, by controlling the amplitude/wavelength ratio of the barrier motion.

If the area of the liquid surface is changed by an amount ΔS, the surface stress, $\Delta\gamma$, can be written as the sum of the elastic and viscous contributions (the latter reflects the effect of any relaxation process in or near the surface) [30]

$$\Delta\gamma = \varepsilon_{d}\mathrm{d}\ln S + \eta_{d}\frac{\mathrm{d}\ln S}{\mathrm{d}t}. \tag{2.8}$$

Here ε_{d} and η_{d} are the surface dilational elasticity and viscosity, respectively.

If the amplitude and rate of the surface deformations are small enough, the coefficients ε_{d} and η_{d} in (2.8) will be constant for a given experiment.

Comparison of Eqs. (2.7) and (2.8) shows that ε incorporates both the elastic and the viscous contributions:

$$\varepsilon = \varepsilon_{d} + \eta_{d}\frac{\mathrm{d}}{\mathrm{d}t}. \tag{2.9}$$

The operator $\mathrm{d}/\mathrm{d}t$ in Eq. (2.9) acts on the horizontal surface displacement, which in an oscillatory experiment is conveniently described as proportional to $\exp(i\omega t)$ (ω is the frequency of the surface excitations). Then one can write

$$\Delta\ln S \sim \exp(i\omega t). \tag{2.10}$$

Hence the operator $\mathrm{d}/\mathrm{d}t$ in Eq. (2.9) is equivalent to a multiplication by $i\omega$, and we will obtain

$$\varepsilon = \varepsilon_{d} + i\omega\eta_{d}. \tag{2.11}$$

Thus in the case of small periodic deformations, the surface Gibbs (dilational) elasticity ε is a complex number, the real part of which equals the surface dilational elasticity ε_{d} and the imaginary part of which is proportional to the surface dilational viscosity η_{d}.

Both surface dilational moduli ε_{d} and η_{d} depend on the frequency ω. This is due to the adsorption–desorption process associated with the monolayer expansion or compression during the propagation of the surface wave (e.g. [30]).

ε_{d} and η_{d} also depend upon the surfactant concentration in the solution, c, (see, for example, [30]). A typical schematic concentration dependence of the surface viscoelastic parameters (ε_{d} or η_{d}), obtained by the just-mentioned surface-waves technique, is shown in Fig. 2.8. For quite low surfactant concentrations ($c \ll$ CMC), the surface viscoelasticity increases as c grows. This reflects the increase in the degree of ordering of the surfactant molecules in the surface monolayer with growing c. When c is a little bit lower than CMC, the surface viscoelasticity reaches its maximal value, which corresponds to the highest molecular ordering in the surfactant monolayer. With c increasing further, the surface viscoelasticity diminishes. This is due to the increasing penetration of the surfactant molecules from the surface monolayer into the bulk of the solution.

The viscoelastic parameters, ε_{d} and η_{d}, have also been evaluated from the data on the thinning rate of some dilute soap films (see Chapter 4).

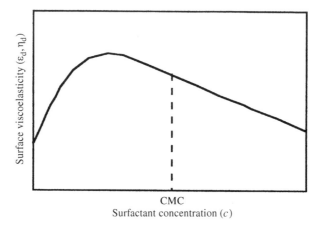

Fig. 2.8 Schematic dependence of the surface dilational viscoelasticity (ε_d or η_d) upon the surfactant concentration, c, for the surfactant solution–air interface (reprinted with permission from Reference [31] Copyright 1988 American Chemical Society)

2.6 PRESSURE INSIDE LIQUID FILMS

2.6.1 CAPILLARY PRESSURE

Consider the profile of a horizontally placed free-standing liquid film stretched over a solid support, Fig. 2.9. The sides of such a film are approximately plane parallel in its central portion, while near the solid support walls the film interfaces are curved due to wetting, i.e. a fluid meniscus (or Plateau's border) with a greater thickness is formed.

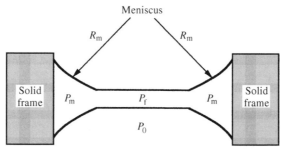

Fig. 2.9 Profile view of a liquid film stretched over a solid support

The interfacial pressure, P_c, which is due to the curvature of the film surface is called the capillary or Laplace pressure and is given by the equation (see, for example, [23])

$$P_c = \frac{\gamma}{R}.$$

(2.12)

Here γ is the surface tension for the liquid of the film and R is the radius of curvature of the film surface.

Since the curvature radius of a quasi-flat central portion of the film $R_f \to \infty$, its contribution to the capillary pressure will be negligibly small, and P_c will be determined only by the curvature radius of the film meniscus R_m: i.e. $R = R_m$ in Eq. (2.12).

2.6.2 DISJOINING PRESSURE

When the thickness of a free-standing film is quite low (of the order of 1000 Å), the molecular interactions between the two interfacial layers can overlap, see Fig. 2.3(b). This leads to excess pressure Π, acting normal to the film interfaces. Π is a function of the film thickness h and can be either positive (determined by repulsive or disjoining forces), or negative (due to attractive or conjoining forces). The first case is most often observed experimentally, so Π is traditionally called the 'disjoining pressure' as was initially proposed by Derjaguin and Obuchov in 1936 [32].

Π is connected with the free energy Φ of a liquid film (see Chapter 1) by the following natural expression (e.g. [33]):

$$\Pi(h) = -\frac{\partial \Phi(h)}{\partial h}.$$

(2.13)

Two main types of disjoining pressure components are present in any liquid film (amorphous or liquid crystalline) film.

(i) The electrostatic disjoining pressure, Π_e, which is due to the repulsive electrostatic forces acting between the double ion layers at the film surface. For a liquid non-electrolyte film, Π_e is given by the following equation (e.g., [33])

$$\Pi_e(h) = 2\varepsilon\varepsilon_0 \left(\frac{k_B T}{ze}\right)^2 K^2.$$

(2.14)

Here ε and ε_0 are, respectively, the dielectric constants of the film substance and of a vacuum, k_B is the Boltzmann constant, T is the absolute temperature, z is the valency of the film substance, e the electron charge, and $K^2 = (z^2 e^2 \rho_0)/(2\varepsilon\varepsilon_0 k_B T)$ (ρ_0 is the density of ions of valency z in the central plane of the film).

(ii) The van der Waals disjoining pressure, Π_{vdW}, which is due, most often, to attractive van der Waals (or dispersive) forces caused by the fluctuations of molecular dipoles of the film medium. In the non-retarded regime Π_{vdW} is given by the following expression (e.g. [33]):

$$\Pi_{vdW} = -\frac{A}{6\pi h^3},\qquad(2.15)$$

where A is the Hamaker constant.

For thin films (h of the order of 10 Å) Π_e can reach values of the order of 10 atm, while Π_{vdW} only reaches values of the order of several tenths of an atm.

A schematic representation of a typical $\Pi(h)$ dependence is shown in Fig. 2.10.

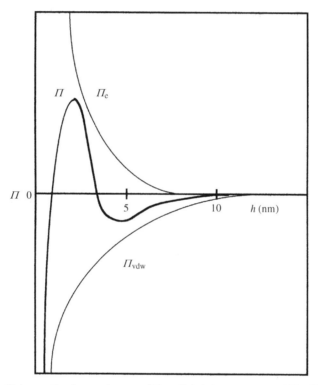

Fig. 2.10   Schematic dependence of the disjoining pressure, $\Pi = \Pi_{el} + \Pi_{vdW}$ (Π_e and Π_{vdW} are, respectively, the electrostatic and the van der Waals components of Π) upon the liquid film thickness, h (Reproduced by permission of Academic Press from Reference [33])

2.6.3 DISJOINING PRESSURE IN LIQUID CRYSTALLINE FILMS

Some specific additional contributions to the disjoining pressure may appear in liquid crystalline films. These are, for example, [32–36]:

(i) elastic disjoining pressure, which is due to the elastic torques caused by deformations of the director field inside the film.

(ii) structural disjoining pressure, which is due to the so-called structural forces caused by the layered molecular ordering inside some mesophase films.

Let us analyse these additional components of Π in more detail.

In a liquid crystalline film bounded by solid substrates, or placed at the surface of some liquid or solid, the elastic deformations inside the film are often due to different boundary conditions at its two surfaces. For a FSLC film different boundary conditions and hence the elastic distortions inside the film also may be realised, for instance, due to local changes in the film thickness (dimples), varied concentrations of the adsorbed ions or a different character of the surface hydrodynamics at the film interfaces. Using the approach of Reference [36], let us calculate the elastic contribution to the disjoining pressure in a FSLC film. Consider an easy idealised example: a nematic free-standing film, with planar and homeotropic boundary conditions at its lower and upper interfaces respectively, Fig. 2.11. The director orientation in such a film is usually called homeo-planar or hybrid.

The free energy Φ of this film in the one-constant elastic approximation can be written as (see Eq. 1.15)

$$\Phi = \Phi_0 + \gamma_0 + \gamma_h + \frac{K}{2} \int_0^h \left(\frac{d\theta}{dz}\right)^2 dz - \frac{W_0}{2} \sin^2 \theta_0 + \frac{W_h}{2} \sin^2 \theta_h. \qquad (2.16)$$

Here γ is the surface tension, h is the film thickness, θ is the angle between the director and the normal to the film surface, $\theta_0 = \theta(0)$, $\theta_h = \theta(h)$, W is the anchoring energy, and indexes '0' and 'h' correspond to the lower and the upper interfaces of the film, respectively.

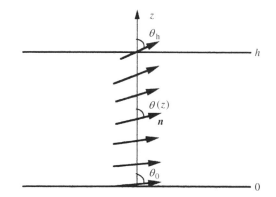

Fig. 2.11 Director distribution in a free-standing nematic film with planar and homeotropic boundary conditions at its lower and upper interfaces, respectively

Note that the surface angles θ_0 and θ_h in Eq.(2.16) are the functions of h: $\theta_0 \Rightarrow 0$, $\theta_h \Rightarrow \pi/2$ for $h \Rightarrow \infty$.

Minimisation of the functional (2.16) with respect to the function $\theta(z)$ gives the following Laplace–Euler expression for the equilibrium distribution of the director and the boundary conditions (the balance between surface and bulk torques):

$$K \frac{d^2\theta}{dz^2} = 0, \tag{2.17}$$

$$K\left(\frac{d\theta}{dz}\right)_{z=0} + \frac{W_0}{2} \sin 2\theta_0 = 0, \tag{2.18'}$$

$$K\left(\frac{d\theta}{dz}\right)_{z=h} + \frac{W_h}{2} \sin 2\theta_h = 0. \tag{2.18''}$$

Eq. (2.17) has the following first integral:

$$\frac{d\theta}{dz} = \frac{\theta_h - \theta_0}{h_{11}}. \tag{2.19}$$

Substituting (2.19) into (2.16), (2.18'), (2.18''), we will obtain

$$\Phi = \Phi_0 + \gamma_0 = \gamma_h + \frac{K(\theta_h - \theta_0)^2}{2 \quad h} - \frac{W_0}{2} \sin^2 \theta_0 + \frac{W_h}{2} \sin^2 \theta_h, \tag{2.20}$$

$$\frac{K(\theta_h - \theta_0)}{h} + \frac{W_0}{2} \sin 2\theta_0 = 0, \tag{2.21'}$$

$$\frac{K(\theta_h - \theta_0)}{h} + \frac{W_h}{2} \sin 2\theta_h = 0. \tag{2.21''}$$

Combining (2.21') and (2.21''), we also obtain the relationship between the anchoring energies and the director surface tilt angles:

$$W_0 \sin 2\theta_0 - W_h \sin 2\theta_h = 0. \tag{2.22}$$

The elastic contribution, Π_{el}, to the disjoining pressure can be calculated from Eq. (2.13), using (2.20), (2.21') and (2.21''):

$$\Pi_{el} = \frac{K(\theta_h - \theta_0)^2}{2 \quad h^2}. \tag{2.23}$$

It is obvious from Eq. (2.23) that the elastic disjoining pressure is positive, i.e. it corresponds to the repulsive interaction.

Since the terms W_0 and θ_0 in (2.23) are not independent, it is not possible to derive an analytical expression for Π_{el} as a function of h. However, a Π_{el} versus h curve can be obtained in the following way: for each value of θ_h Eq. (2.22) gives θ_0, then Eqs. (2.21') and (2.21'') give h, and Π_{el} is eventually obtained from (2.23). Fig. 2.12 shows the result of such calculations for fixed typical values of the anchoring energies at the lower and upper film interfaces. It is obvious from this Figure that the elastic disjoining pressure is significant at film thicknesses of the order of 0.1 μm,

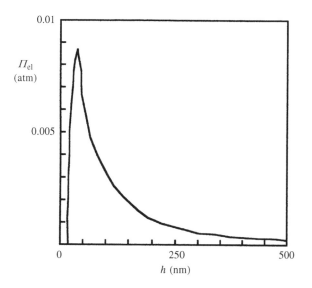

Fig. 2.12 Computed values of the elastic disjoining pressure, Π_{el}, in a nematic film with homeo-planar boundary conditions versus the film thickness, h: elastic constant $K = 7 \times 10^{-7}$ dyne; anchoring energy at the lower film interface, $W_0 = 2.2 \times 10^{-1}$ erg/cm^2; at the upper film interface $W_h = 1.1 \times 10^{-1}$ erg/cm^2 (reprinted from Reference [36] courtesy of Marcel Dekker, Inc.)

where van der Waals and electrostatic components of the disjoining pressure can be neglected (see Fig. 2.10).

The just-obtained theoretical estimations of Π_{el} are in satisfactory agreement with the experimental data on the disjoining pressure of liquid crystalline films made in the Scheludko cell [35, 36] (see also Chapter 3).

Note that we have considered the FSLC film with only splay–bend elastic deformations. However, the same equations can be applied for the case of twist distortions.

The structural oscillating component of the disjoining pressure is observed experimentally by the Israelachvili force measurement technique and in the Scheludko and Mysels cells (see Chapter 3) for thin ($h < 1000$ Å) liquid crystalline soap polymer and even for some amorphous liquid films (see, for example, [33, 34, 36–38]). The amplitude of the disjoining pressure oscillations, Π_s, has very low values; for example, for micellar soap films it is of the order of 10^{-4} atm for $h =$ several tens of nm [37]. The Π_s term diminishes exponentially with growing h (e.g. [33, 39]):

$$\Pi_s(h) = k_s \exp\left(-h/l_s \cos(\pi\beta h)\right), \qquad (2.24)$$

where k_s and β are empirical constants and l_s is the empirical correlation length.

The described oscillations are due to layer by layer film thinning (stratification) during the measurements. For more detail on the stratification phenomenon see Chapter 4.

2.6.4 THE PRESSURE BALANCE IN A FREE-STANDING FILM

Taking into account the facts that the gas pressure P_0 around the film is the same everywhere and that the gravity effects for the horizontally placed and quite thin ($h < 1$ μm) film can be neglected, the following pressure balance equations can be written (see Fig. 2.9):

$$P_f = P_0 - \Pi \tag{2.25'}$$

$$P_m = P_0 - P_c. \tag{2.25''}$$

Here P_f and P_m are, respectively, liquid pressures inside the central portion of the film and inside the film meniscus.

The pressure gradient $\Delta P = P_f - P_m$ is usually called the 'driving pressure'. Indeed, when $\Delta P > 0$, the liquid drains from the central portion of the film to the film meniscus, which causes film thinning.

The disjoining and capillary pressures, acting together, determine the equilibrium thickness h_{eq} of the liquid film. Indeed, at equilibrium $\Delta P = 0$ and hence $\Pi = P_c$.

REFERENCES

1. Sonin A.A., 1995, *The Surface Physics of Liquid Crystals*, OPA-Gordon and Breach, Amsterdam.
2. Sheng P., 1982, *Phys. Rev. A*, **26**, 1610.
3. Faetti S., Palleschi V., 1984, *Phys. Rev. A*, **30**, 3241.
4. de Gennes P.G., 1974, *The Physics of Liquid Crystals*, Clarendon Press, Oxford.
5. de Gennes P.G., Prost J., 1993, *The Physics of Liquid Crystals*, 2nd edn., Clarendon Press, Oxford.
6. *Thin Liquid Films* (ed. I.B. Ivanov), *Surfactant Science Ser.*, Marcel Dekker, New York, 1988, vol. 29.
7. Nikolov A.D., Kralchevsky P.A., Ivanov I.B., Wasan D.T., 1989, *J. Colloid Interface Sci.*, **133**, 13.
8. Kralchevsky P.A., Nikolov A.D., Wasan D.T., Ivanov I.B., 1990, *Langmuir*, **6**, 1180.
9. Langevin D., Sonin A.A., 1994, *Adv. Colloid Interface Sci.*, **51**, 1.
10. Adamson A.W., 1976, *Physical Chemistry of Surfaces*, 3rd edn. Wiley, New York.
11. Brochard F., de Gennes P.G., Pfeuty P., 1976, *J. Phys. (France)*, **37**, 1099.
12. David F., Leibler S., 1991, *J. Phys. (France) II*, **1**, 959.
13. Pieranski P., Beliard L., Tournellec J.-Ph., Leoncini X. *et al.*, 1993, *Physica A*, **194**, 364.
14. Rapini A., Papoular M.J., 1969, *J. Phys. (France) Colloq.*, **30**, C4–54.
15. Blinov L.M., Kabayenkov A.Yu., Sonin A.A., 1989, *Liq. Cryst.*, **5**, 645.
16. Stoebe T., Mach P., Huang C.C., 1994, *Phys. Rev. E*, **49**, R3587.
17. Mach P., Grantz S., Debe D.A., Stoebe T., Huang C.C., 1995, *J. Phys. (France) II*, **5**, 217.
18. Stoebe T., Mach P., Grantz S., Huang C.C., 1996, *Phys. Rev. E*, **53**, 1662.

19. Mach P., Huang C.C., Nguyen H.T., 1998, *Phys. Rev. Lett.*, **80**, 732.
20. Miyano K., 1982, *Phys. Rev. A*, **26**, 1820.
21. Stannarius R., Cramer Ch., 1997, *Liquid Crystals*, **23**, 371.
22. Stannarius R., Cramer Ch., 1998, *Europhys. Lett.*, **42**, 43.
23. Rosen M.J., 1989, *Surfactants and Interfacial Phenomena*, 2nd edn. Wiley, New York, p. 276.
24. Langevin D., Meunier J., 1994, in *Micelles, Membranes, Microemulsions, and Mono-layers*, (eds. W.M. Gelbart, A. Ben-Shaul, D. Roux), *Partially Ordered Systems Ser.*, Springer, New York, p. 485.
25. Kazakov N.B., Kaznacheev A.V., Sonin A.S., 1991, *Zh. Fiz. Khim.*, **65**, 783.
26. Gibbs J.W., 1928, *Collected Works*, Longmans Green, New York, vol. 1, p. 302.
27. Rayleigh J.W.S., 1890, *Proc. R. Soc. London, Ser. A*, **48**, 127.
28. Marangoni G., 1872, *Nuovo Cimento*, **2**, 239.
29. Plateau J.A.F., 1869, *Philos. Mag.*, Ser. 4, **38**, 445.
30. Lucassen-Reynders E.H., 1981, in *Anionic Surfactants. Physical Chemistry of Surfactant Action*, (ed. E.H. Lucassen-Reynders), *Surfactant Science Ser.*, Marcel Dekker, New York, vol. 11, p. 173.
31. Stenvot C., Langevin D., 1988, *Langmuir*, **4**, 1179.
32. Derjaguin B.V., Obuchov E., 1936, *Acta Physicochim. URSS*, **5**, 1.
33. Israelachvili J.N., 1992, *Intermolecular and Surface Forces*, 2nd edn., Academic Press, London.
34. Horn R.G., Israelachvili J.N., Perez E., 1981, *J. Phys. (France)*, **42**, 39.
35. Proust J.E., Ter-Minassian-Saraga L., 1979, *J. Phys. (France) Colloq.*, **40**, C3–490.
36. Perez E., Proust J.E., Ter-Minassian-Saraga L., in Ref. 6, p. 891.
37. Bergeron V., Radke C.J., 1992, *Langmuir*, **8**, 3020.
38. Asnacios A., Espert A., Colin A., Langevin D., 1997, *Phys. Rev. Lett.*, **78**, 4974.
39. Bergeron V., Jiminez-Laguna A.I., Radke C.J., 1992, *Langmuir*, **8**, 3027.

3

Experimental Techniques

In this chapter we will consider the main experimental methods used for preparation and investigation of freely suspended liquid films.

3.1 FILM PREPARATION

According to [1], we will divide all the existing methods of flat liquid film preparation into two groups.

1. A particle (bubble, drop or solid sphere) is released into a liquid and allowed to approach the liquid interface. As a result, a thin film is formed between the particle and the interface.
2. 'Supported' liquid films are formed on the frames, in the capillaries, etc.

Probably, Derjaguin and Kussakov (1939) were the first to apply the former group of methods to investigate liquid film drainage at a solid substrate [2]. They studied the approach of small air bubbles in water to flat sheets of freshly cleaved mica or freshly blown glass. The same method was used later on to obtain free-standing liquid films: for two gas bubbles coalescing in a liquid [3] and for a gas bubble coalescing at a liquid–air interface [4] (Fig. 3.1).

The techniques from the second group (namely preparation of films on solid supporting frames of different forms) were utilised even in early work by Perrin and Friedel (see section 1.1) on free-standing films. A frame should be plunged into a liquid, from which the film is to be prepared, and then slowly pulled out from it. Exactly like this, with special small, plastic frames, children produce their soap bubbles. Another possibility is to stretch a film over a frame by means of a blade, for example.

One of the supporting frames most commonly used for the preparation of freely suspended liquid films in modern studies is represented in Fig. 3.2. This is a rectangular-form frame consisting of two cylindrical metal wires (1 and 2) of diameter of the order of $10\,\mu m$ and two cylindrical nylon wires (3 and 4) of the same diameter (e.g. [5, 6]). The inter-wire distances a and b can be gradually changed by displacing the wires 1 and 3 (for instance, by means of a micrometric

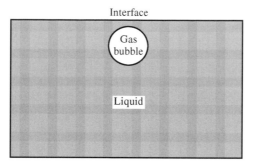

Fig. 3.1 Formation of a flat, free-standing film between a gas bubble and a liquid–air interface (from Reference [4]. Reproduced by permission of Academic Press)

screw connected with an electric motor). This permits one to change the film thickness. The external voltage, U, can be applied to the conducting wires 1 and 2.

Such frames are convenient for studying, for example, the spontaneous or field-induced orientational transitions and EHD (electrohydrodynamic) instabilities in FSLC films (see Chapters 5 and 6).

Supporting frames of other geometries (for instance, circular-form wire loops) are also used (e.g. [7]).

The technique, combining a supporting frame and a capillary, was proposed by Scheludko (1957) [8] [Fig. 3.3(a)] see also Section 1.1. His cell, for preparation of liquid free-standing films, consists of a cylindrical-form glass supporting frame (1) of a diameter D of the order of 1 mm, attached by a glass capillary tube (2) to a glass reservoir (3). The air pressure, P, inside the reservoir can be changed, for instance, by moving a mercury piston (4) with the help of a micrometric screw (5). A liquid drop is sucked out by lowering P, and a flat film, surrounded by a meniscus, is thus formed in the centre of the frame (1), see Fig. 3.3(a). This technique allows one to prepare a free-standing liquid film of a given diameter, $D_f \approx 0.1$–1 mm, controlling the driving pressure, ΔP (see Chapter 2). ΔP (and hence the disjoining pressure, Π)

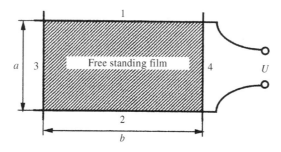

Fig. 3.2 Rectangular-form support frame for preparation of freely suspended liquid films: (1) and (2) metal wires; (3) and (4) nylon threads (adapted from References [5, 6]. Copyright 1993 American Institute of Physics)

can be measured by means of a manometer or an electronic membrane sensor. The Scheludko method permits one to work only in the limit of quite low driving pressures (lower than the capillary pressure in the tube 2). Above this value the air is drawn through the tube.

A modified Scheludko cell allowing one to deal with higher driving pressures (up to several atm) was proposed by Mysels (1964) [9] and perfected in several later works [e.g., [10–13], Fig. 3.3(b)]. Here, instead of a cylindrical supporting frame, a plate of porous glass (1) with a cylindrical hole drilled through it (2) is used. A film is formed in the hole (2) by sucking out the liquid through the porous glass by the attached capillary tube (3). This technique permits one to form free-standing films of a smaller diameter ($D_f \approx 0.1$ mm) than those obtained by the classical Scheludko method.

In some cases, to prevent the evaporation for free-standing films prepared from surfactant solutions, a supporting frame with a film may be placed into a hermetic box containing the film liquid poured in at the bottom (e.g. [12, 13]).

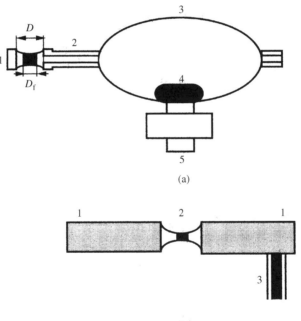

Fig. 3.3 The techniques for the free-standing films preparation (reprinted with permission from Reference [9]. Copyright (1964) American Chemical Society). (a) The Scheludko method: (1) cylindrical-form glass supporting frame; (2) capillary tube; (3) glass reservoir; (4) mercury piston; (5) micrometric screw; D and D_f are, respectively, the diameters of the supporting frame and of the obtained liquid film. (b) The Mysels method: (1) porous glass plate; (2) cylindrical hole; (3) capillary tube

3.2 FILM THICKNESS MEASUREMENTS

The thickness h of a free-standing film is an important parameter for the quantitative characterisation of its thermodynamic behaviour. A sufficiently large number of experimental methods exist allowing one to determine h. These are, for example the evaluation of the film thickness from the capacitance, and resistance measurements (e.g. [14–17]); from the measurements of the period of the EHD instability ([5], see Chapter 6, Eq. 6.2). All these techniques are not very precise, since only the average film thickness can be determined. Moreover, the presence of metal electrodes is always required.

The most widespread methods of measuring h are based on optical phenomena: the interference of light reflected from both interfaces of a film. The experimental technique using this principle was initially utilised in 1918, for soap films, by J. Perrin (see also paragraph 1.1) and since has been perfected and applied to different free-standing films. In modern practice, to measure h in a small portion of the film, a focused laser beam is usually used.

If the film is not homogeneous in its thickness, the radius–vector distribution, $h(r)$, can be found simply from observations of the Newton's interference rings under a microscope (e.g. [2, 4]).

The time evolution, $h(t)$, for a thinning liquid film can be measured by the interferometric technique proposed by Scheludko in [8, 18]. An oscillating time dependence of the light intensity I, reflected from some fixed area of the film, is measured (Fig. 3.4); and h is calculated later on using the formula for I reflected from a sheet with plane-parallel surfaces [18, 19]:

$$h = \frac{\lambda}{2\pi n} \arcsin \left[\frac{\Delta}{1 + 4R(1 - \Delta)/(1 - R^2)} \right]^{1/2}. \qquad (3.1)$$

Here λ is the wavelength of the incident light, n is the refractive index of the film liquid, $\Delta = (I - I_{min})/(I_{max} - I_{min})$, I is the measured intensity of the reflected light, I_{min} and I_{max} are its minimal and maximal values, and $R = (n - 1)^2/(n + 1)^2$.

In the case of thick films (when $I(t)$ exhibits many extreme values; see Fig. 3.4), it is convenient to use a simplified Eq. (3.1)

$$h = \frac{\lambda}{2n} k, \qquad (3.2)$$

where $k = 1, 2, 3, \ldots$; $I = I_{max}$ or $I = I_{min}$.

If such a liquid film ruptures at some thickness, h_c, higher than the thickness of the Newton black film, h_b, ($h_b \approx$ several tens of Å), it is not possible to measure the precise value of h at a given t, and only the relative plot $h/h_0(t)$ (where h_0 is the initial film thickness) can be obtained. However, if the film reaches the black Newton state, its thickness, h_b, can then be known with the precision of about ten Å, which will be the error value for the $h(t)$ curve calculated from Eqs. (3.1) and (3.2); see Fig. 3.4.

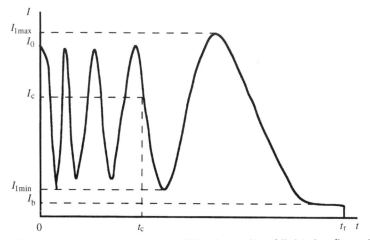

Fig. 3.4 Schematic time dependence of the intensity of light, I, reflected from a thinning free-standing liquid film: I_0 corresponds to the initial film thickness, h_0, at $t = 0$; I_c corresponds to some thickness, h_c, at which the film may rupture; I_{1min} and I_{1max} are, respectively, the intensities of the last minimum and maximum (h_{1min} and h_{1max} are corresponding film thicknesses); I_b is the intensity of light reflected from the Newton black film (h_b is the Newton black film thickness); t_r relates to the moment of the black film rupture

To overcome the just-mentioned difficulty, interference filters are used in some devices. They allow one to use two different wavelengths, λ, in order to facilitate the determination of k in Eq. (3.2) and, hence, the precise value of h [13].

If the thickness of a free-standing film does not change with time, it is more convenient to use other variation of the interferometric method (e.g. [20–23]).

As an example consider a method, proposed in [21], based on the use of the Pohl interferometer (Fig. 3.5). A cell containing a liquid film is fixed on a rotation table. A laser beam is focused on the film with the help of a lens. The reflected light is detected on a screen in the form of parallel equidistant interference fringes. The spacing, Δx, between the neighbouring fringes gives the film thickness, h:

$$h = \frac{z\lambda(n^2 - \sin^2\theta)^{1/2}}{2\Delta x \sin\theta \cos\theta}. \tag{3.3}$$

Here z is the cell-to-screen distance (normally z is of the order of 1 m); λ is the laser wavelength; n is the film material refractive index at the laser wavelength; θ is the angle of incidence of the laser beam with respect to the film normal (in practice it is often convenient to use $\theta = 45°$).

Note that the value of n in Eqs. (3.1)–(3.3) depends on the orientation of the liquid crystalline director. For example, for a nematic film with a homeotropic orientation of the director $n = n_o$, where n_o is the ordinary refractive index; for planar or twisted orientations, $n = n_e$, where n_e is the extraordinary refractive index; for tilted orientations (or for a nematic film in an isotropic phase), $n = 1/3(n_e + 2n_o)$.

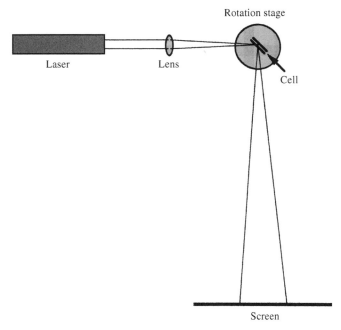

Fig. 3.5 Measurement of the free-standing film thickness by means of the Pohl interferometer (from Reference [21] by permission of Gordon and Breach Science Publishers)

The latter method just described was first used to evaluate the thickness of a liquid crystalline film, bounded by solid plates [21], and later on was successfully utilised for measurements of FSLC film thicknesses (e.g. [25]).

3.3 OPTICAL MICROSCOPY

This quite straightforward experimental technique may give a lot of useful information about FSLC films. For example, the director distribution in different FSLC films may be characterised directly by polarisation microscope observations in transmitted or reflected light or by forms of conoscopic images (e.g. [24–26]). The surface two-dimensional hydrodynamics (dynamics of the stratification domains in soap films, of the EHD domain patterns in thermotropic liquid crystalline films, etc.) and diffusion can also be studied by this method (e.g. [5, 6, 13, 27–29]). To make a two-dimensional motion clearly visible, one often places different small solid particles at the free-standing film interface (e.g. [27, 28]), or dissolves fluorescent compounds in film-forming substances (e.g. [29]).

Computer treatment and storing of the microscopic images are often utilised in the most recent studies (e.g. [24, 29]).

3.4 POLARISATION-OPTICAL METHOD

This technique allows one to determine the birefringence of a free-standing film, which, in principle, makes it possible to calculate such important properties of the film as the mean angle of the director orientation, order parameters, the surface tension, etc. (see, for example, [30–32]).

A schematic representation of the polarisation-optical set-up is shown in Fig. 3.6. The beam of a He–Ne laser, polarised by the first nicol (polariser), comes through a FSLC film and, after that, through the second nicol (analyser). The final light signal is registered by a photodiode or photomultiplier. To increase the sensitivity, lock-in amplification and synchronous detection are often used (for more details see, for example, [30, 31]).

The intensity, I, of the laser light transmitted through two nicols and a FSLC film can be calculated by the following formula (e.g. [30, 31]):

$$ I = I_0 \sin^2 2\beta \sin^2 \frac{\Delta\Phi}{2}. \tag{3.4} $$

Here I_0 is the light intensity for maximal transmission, β is the angle between the polariser axis and the liquid crystalline director (in practice, it is often convenient to use $\beta = 45°$), and $\Delta\Phi$ is the phase delay for the transmitted light.

It is obvious from Eq. (3.4) that, if the director orientation in the film (i.e. $\Delta\Phi$) alters under some external action (e.g. electric or magnetic fields, temperature, etc.), I will be changed in an oscillating fashion (as in Fig. 3.4).

Consider, for instance, a thermotropic nematic free-standing film (∞/mm symmetry). The optical indicatrice in this case is an ellipsoid of circular cross-section, whose axes are the principal refractive indices, n_o and n_e, (Fig. 3.7). Let, due to some external action (e.g. of an electric or magnetic field), the long axis of this ellipsoid be inclined by some angle θ' to its initial orientation (along 00-axis). If the incident light is parallel to the z-axis (i.e. normal to the film interfaces), the birefringence value, Δn, will be given as the difference between the lengths of the

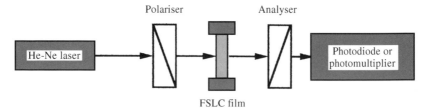

Fig. 3.6 Schematic view of the polarisation-optical experimental set-up (adapted from Reference [32]. Reprinted by permission of Gordon and Breach Publishers)

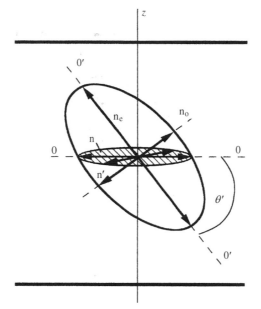

Fig. 3.7 The optical indicatrice of a nematic liquid crystal (from Reference [32] reprinted by permission of Gordon and Breach Publishers)

axes n and n' of the cross-section of the ellipsoid perpendicular to z. Thus we get (e.g. [30, 31])

$$\Delta n(z) = n - n' = \frac{n_o n_e}{[n_e^2 \sin^2 \theta'(z) + n_o^2 \cos^2 \theta'(z)]^{1/2}} - n_o \qquad (3.5)$$

Or, for small values of θ' ($\theta' \ll 1$)

$$\Delta n(z) \approx \frac{1}{2} n_o \left(1 - \frac{n_o^2}{n_e^2} \right) \theta'^2(z). \qquad (3.6)$$

Taking into account that $\Delta\Phi = 2\pi/\lambda \langle \Delta n \rangle$ (where λ is a wavelength of the incident light, $\langle \Delta n \rangle$ is the average over the film thickness birefringence), and using (3.6), it is easy to obtain the general relation between the experimentally determined value of $\Delta\Phi$ (see Eq. 3.4) and the mean square of the director tilt angle $\langle \theta'^2 \rangle$:

$$\Delta\Phi = \frac{\pi}{\lambda} n_o \left(1 - \frac{n_o^2}{n_e^2} \right) < \theta'^2. \qquad (3.7)$$

3.5 ELLIPSOMETRY

The ellipsometric technique is based upon the representation of the electric vector, E, of elliptically polarised light, propagated in some medium along the z co-ordinate axis, as the sum of the two following mutually perpendicular wave-components, respectively parallel to the co-ordinate axes x and y:

$$E_x = A_x \cos(\omega t + \Theta_x), \qquad (3.8')$$

$$E_y = A_y \cos(\omega t + \Theta_y). \qquad (3.8'')$$

Here A_x and A_y are the amplitudes of these electric waves, ω is their angular frequency, and θ_x and θ_y are their initial phases.

In practice, the two ellipsometric parameters, Δ and φ, connected with the optical properties of the medium, are usually measured by means of special optical devices—ellipsometers (for more information see, for example, [33–35]):

$$\Delta = \Theta_x - \Theta_y, \qquad (3.9)$$

$$\operatorname{tg} \varphi = \frac{A_x}{A_y}. \qquad (3.10)$$

The ellipsometric method is successfully utilised for determination of thicknesses and optical constants of different thin films deposited on solid substrates (e.g. [33–35]). However, it can also be applied to FSLC films. In References [36–39], for example, ellipsometry was used for the characterisation of different phases and for the determination of some physical parameters (tilt angle, spontaneous polarisation) in free-standing thermotropic smectic films.

3.6 CALORIMETRY

Calorimetric measurements of the heat capacity, C_p, is an effective tool to study FSLC films, and their phase transitions, in particular, since in phase transition points C_p exhibits maxima.

A schematic representation of the calorimetric system, used for studying phase transfers in thermotropic smectic free-standing films (see, for example, [40–43]), is shown in Fig. 3.8. Mechanically modulated radiation from He–Ne laser is focused at the film surface. This radiation is absorbed by a film formed in a 1-cm-diameter hole in a metallic frame and placed in a hermetic thermo-stabilised cell. The cell is filled with argon gas at a pressure of 0.5 atm. The argon atmosphere serves for better temperature exchange between the film interface and the first thermocouple, placed just below the film at a distance of about 0.25 μm from its lower surface. The second thermocouple is placed lower, in order to measure the average sample temperature. The difference of signals from two thermocouples multiplied by a lock-in amplifier, operating in the differential mode, gives the magnitude of the film temperature oscillation. This temperature oscillation is directly related to C_p.

Note that because of the fairly small thermal diffusion length of the argon exchange gas (≈ 0.6 mm) at the radiation wavelength used (3.39 μm), only a very small portion (≈ 0.6 mm in diameter) of the film is actually probed by the calorimetric system. This means that for a two-molecular-layer film for instance, the amount of the sample contributing to the heat capacity measurement is less than 20 ng in weight! So, a great sensitivity in the detection of the thermocouple probe signals is needed. The system resolution achieved at the moment is better than a few parts in 10^5 for both probes [41].

The described experimental set-up also allows parallel measurements of the light reflectivity (see also Section 3.8.1) from the film surfaces (using another He–Ne laser as a light source), in order to determine the film thickness and the in-plane molecular density for smectic films [41–43] (see Fig. 3.8).

3.7 SURFACE TENSION MEASUREMENTS

The traditional Wilhelmy technique (e.g. [44]) is not applicable for evaluation of the surface tension, γ_f, of freely suspended films, due to the large dimensions of the Wilhelmy ring or plate with respect to the film thickness.

An extravagant and easy method for the determination of γ_f has been elaborated in [45], Fig. 3.9. The rectangular slot of a metallic supporting frame for the formation of a freely suspended liquid film is divided into two equal parts by a very flexible, uniform thickness (several tens of μm), silk surgical string, fixed to a cell at its upper end and stretched by a mass m, hang at its lower end. When there is no film in the

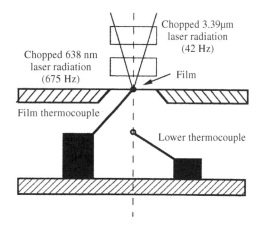

Fig. 3.8 Schematic view of a highly sensitive calorimeter for studying freely suspended liquid films (reproduced from Reference [40] with permission. Copyright the American Physical Society 1986)

frame, the string is straight. When a liquid film is formed in one part of the frame (between the string and the cell edge), the string will be deformed into an arc, due to the action of the film surface tension. It was shown that the friction between the supporting frame and the string is negligible. Then, a simple equation for γ_f may be obtained from the force balance [45]:

$$\gamma_f = mg/2r, \tag{3.11}$$

where r is the arc curvature radius.

The modification of the just-described method, where a mylar trapeze is used instead of a string, is described in [46]. This trapeze is attached to a small metallic pendulum and a mirror. The film is stretched between the trapeze side and the cell edge. Under the action of the film surface tension γ_f, the trapeze is deflected from its equilibrium position at some angle. This angle (which is connected with γ_f) can be measured by the reflection of the laser beam from the mirror.

The other interesting method to determine γ_f is based on the preparation of spherical bubbles. In this case, the surface tension of the bubble wall (a free-standing film) is given by the following simple relation [47, 48]:

$$\gamma_f = \Delta P\, R/4, \tag{3.12}$$

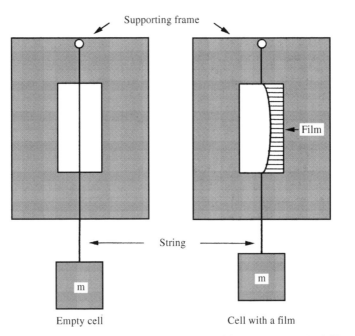

Fig. 3.9 Measurements of the surface tension of freely suspended films by the deformation of a silk string (reproduced with permission from Reference [45]. Copyright 1994 by the American Physical Society)

where ΔP is the difference between the pressures inside and outside the bubble, and R is the bubble radius.

Finally, γ_f can be evaluated from measurements of the frequencies of the FSLC film interfacial vibrations (e.g. [46, 49, 50]). Such vibrations may be induced electrically, by applying an a.c. voltage between a group of cylindrical electrodes, placed just below the film, and the film-containing circular (or rectangular) metallic frame [46, 49] (Fig. 3.10); or, acoustically, with the help of a loud-speaker placed beneath the film [50]. The vibrations are detected using a laser beam reflected from the film surface.

The surface tension γ_f can be calculated from the following formula for the resonance frequencies ($\omega_{n,s}$) of the circular membrane vibrations in a vacuum, obtained by Rayleigh [51]:

$$\omega_{n,s} = (u_{n,s}/R)(\gamma_f/\rho_f)^{1/2}. \tag{3.13}$$

Here $u_{n,s}$ is the sth root of the Bessel function $J_n(r)$, r is the radius-vector, R is the film radius, and $\rho_f \approx \rho h$ is the film (two-dimensional) density (ρ is the film-material bulk density, h is the film thickness).

All the described techniques were successfully applied for measurements of γ_f in some thermotropic smectic films (see also Chapter 2).

As already mentioned in Chapter 2, as far as we know, at the moment, there are no experimental data on the anisotropy of the surface tension (i.e. the anchoring energy,

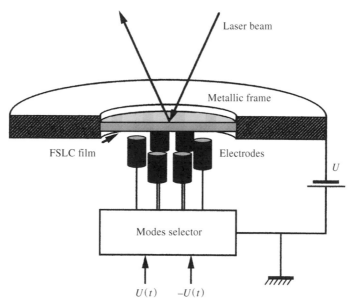

Fig. 3.10 Measurements of the free-standing film surface tension by the vibration method (reprinted from Reference [46]. Copyright 1993 with permission from Elsevier Science)

W_f) for FSLC films. However, W_f can in principle be estimated by one of the traditional methods used for its determination in liquid crystals (see, for example, [32]): for instance, from the threshold of the Fredericks transition.

3.8 STRUCTURAL STUDIES OF FREE-STANDING LIQUID CRYSTALLINE FILMS

The experimental techniques based on the diffraction and reflection of electromagnetic radiation and of elementary particles normally utilised for structural investigations of a condensed matter, are also largely used to study the structure, intermolecular forces and behaviour of fluctuations of FSLC films. Since the structural study experiments often take a considerable time (depending on the technique used), only thermodynamically stable films, which practically do not change their thickness and molecular structure with time, can be investigated. These are, for example, thermotropic smectic and black soap films (see also Chapters 4, 5 and 7). Here we will briefly review the main features of some of the experimental methods used for structural studies of FSLC films.

3.8.1 LIGHT SCATTERING AND REFLECTION

Thermally excited surface corrugations (or fluctuations) of two types, bending and squeezing (Fig. 3.11), are always present at interfaces of any free-standing liquid film. These corrugations have a wavelength of the order of magnitude of the wavelength of visible light, and thus may scatter incident optical radiation. In contrast, their amplitude is very small, of the order of several Å (see, for example, [52]). As the thickness, h, of the free-standing film is decreased, the contribution to

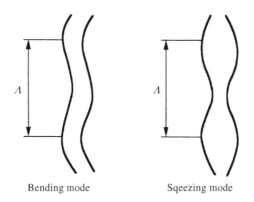

| Bending mode | Sqeezing mode |

Fig. 3.11 Bending and squeezing modes of surface corrugations at the interface of a free-standing liquid film. Amplitudes are highly exaggerated compared with the wavelength Λ (reprinted from Reference [52] courtesy of Marcel Dekker, Inc.)

the surface fluctuations from the squeezing mode becomes smaller, whereas, from the bending mode, it becomes greater [53].

The contribution to the light scattering intensity from these two surface fluctuation modes is essential in lyotropic FSLC films. In free-standing films made of thermotropic liquid crystalline phases, however, an additional prevailing contribution appears. This is the light scattering from thermal orientational fluctuations of the director. Thus in thin (two-dimensional) smectic, freely suspended films, two types of orientational fluctuations are found: bend and splay, which are characterised by the corresponding elastic constants (see also Chapter 1) [54–56].

Two principle varieties of the light scattering method exist. Both of them are used to study FSLC films. These are:

1. classical static light scattering, where the mean over a certain time scattered intensity, $\langle I(t) \rangle$, is measured;
2. dynamic light scattering, where the correlation functions, $\langle I(0)I(t) \rangle$, are determined.

A schematic representation of the static light scattering set-up is shown in Fig. 3.12(a). It consists of a laser, a lens focusing the laser beam, a thermo-stabilised cell containing a liquid film, and a photomultiplier tube, which contains a photomultiplier (PMT) itself, and a detection optical part. The PMT is placed at some angle χ with respect to the incident beam, and serves to detect the light intensity scattered at this angle. The χ value may be changed.

The static light scattering technique allows one to obtain the mean-square amplitudes of the surface and director orientational fluctuations.

A scheme of the dynamic light scattering set-up is shown in Fig. 3.12(b). The basic part of this equipment is the same as for the static light scattering technique, but several new elements are added. The scattered light signal detected by the PMT is treated either by a correlation, or a spectrum analyser. It consists of a fast analogue-to-digital converter and a minicomputer that is programmed to calculate the time auto-correlation function of the PMT signal. This correlation function is proportional to the time auto-correlation function of the fluctuations in the light scattered intensity. The output of the spectrum analyser is proportional to the power spectrum of the scattered intensity fluctuations.

The dynamic light scattering method permits the study of the time evolution of the surface and the director orientational fluctuations, to measure, for example, their relaxation times. Most often both these light scattering techniques are used together.

More details on light scattering may be found, for instance, in the book [57]. Analysis of the light scattering data permits the evaluation of some macroscopic thermodynamic parameters of FSLC film. Thus the values of intermolecular forces (disjoining pressure) and the surface tension can be calculated from the amplitudes of the surface fluctuations (see, for example, [52, 58] and Eq. (4.37)). It also allows one to evaluate the energetic characteristics of thermotropic smectic free-standing films, such as the elastic moduli and viscosities (e.g. [59–64]). In addition,

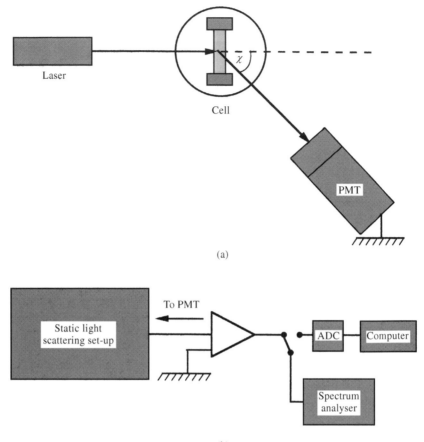

(a)

(b)

Fig. 3.12 Principle schemes of the static (a) and dynamic (b) light scattering techniques

macroscopic polarisation may be determined for ferroelectric smectic films (e.g. [59–61]).

Light scattering was also successfully used for quantitative characterisation of the BO (bond-orientational) order (see also Chapters 1 and 5) in freely suspended films prepared from thermotropic smectics with structured layers (e.g. [65, 66]).

Some applications of the light reflection technique to FSLC films have already been mentioned in Sections 3.2 and 3.6. Using this method one can also study the dynamics of orientational fluctuations in free-standing thermotropic smectic films [67, 68]. The spontaneous polarisation in freely suspended ferroelectric smectic films can be also measured by means of this method; and the relaxation processes in such films submitted to the action of an external electric field can be investigated [69–72].

3.8.2 X-RAY SCATTERING AND REFLECTION

The X-rays methods, applied to FSLC films, may give useful information about their phase metamorphism and molecular organisation. To obtain a sufficient signal-to-noise ratio, it is in practice necessary to use an X-ray beam with an area of the order of $1 \, mm^2$, i.e. to study quite large homogeneous films (with areas of the order of $1 \, cm^2$). It is often rather difficult to prepare homogeneous thermotropic smectic films, since they have a tendency to divide themselves into a number of domains having a uniform director alignment. To increase the degree of the director orientation, a magnetic field is generally used [73, 74].

In practice, two principle geometries of the X-ray technique are used: diffraction and reflection. For low-resolution diffraction studies of FSLC films the standard X-ray diffractometers are utilised (e.g. [75, 76]), while for high-resolution diffraction studies the synchrotron beam lines are used (e.g. [76]). More technical details on the latter method may be found, for instance, in Reference [77].

Study of the X-ray diffraction patterns allows one to determine the symmetry group of a FSLC film, i.e. to identify the film phase. X-ray diffraction has already been used fruitfully to investigate the phase polymorphism in free-standing thermotropic smectic films (see, for example, [78, 79] and Chapter 5). For example X-ray diffraction gives information about the BO ordering in free-standing films, prepared from thermotropic smectics with structured layers. The presence of hexatic BO ordering in the layer planes of such films results in a six-fold modulation of the scattered intensity $I(\chi)$ of the X-ray radiation, which can be described by a Fourier series as (e.g. [73, 74])

$$I(\chi) = I_0 \left\{ \frac{1}{2} + \sum_{n=1}^{\infty} C_{6n} \cos \left[6n(90° - \chi) \right] \right\} = I_{bg}. \tag{3.14}$$

Here χ is the scattering angle, I_0 is the incident radiation intensity, $C_{6n} = \mathrm{Re}\langle Y_6^n \rangle = \mathrm{Re}\langle \exp(i6n\omega) \rangle$ are the $6n$-fold BO order parameters (see Eq. 1.8), and I_{bg} is the background radiation intensity. The BO order parameters can thus be estimated from the data on the scattering radiation intensity.

X-ray diffraction can also be used to reconstruct the FSLC film electron density profiles in the direction perpendicular to the film interface, i.e. to obtain a real atomic and molecular structure of a film (e.g. [80]). However, the X-ray reflection method is more often utilised for these purposes. Thus the analysis of the X-ray reflection patterns was recently applied to structural studies of both thermotropic smectic [81] and black soap [82] films (see also Chapter 7).

Finally, some attempts to apply X-ray diffraction to the study of thermal fluctuations in free-standing smectic A films have been made recently [83, 84].

3.8.3 ELECTRON DIFFRACTION

Because electrons have a wave-like character they may be reflected or refracted in ways similar to light or X-ray radiation. In spite of its quite low resolution with respect to the X-ray methods, electron diffraction may serve as a useful tool for the structural investigations of FSLC films. Indeed, an electron microscope (used as a diffractometer) allows easy focus of an electron beam onto a small spot (with an area of the order of $1\,\mu m^2$) inside a homogeneous domain of a smectic free-standing film. Thus this technique does not demand large homogeneously oriented films, which simplifies considerably the sample preparation (see the previous paragraph).

Electron diffraction was successfully used to study the phase metamorphism and phase transitions in free-standing smectic films (e.g., [85–88]). For instance, the BO order parameter was calculated (using the same approach as for X-ray diffraction, see Eq. 3.14 and Chapter 5) for smectic structural layers [85, 86].

3.8.4 CRITICAL REFLECTION OF NEUTRONS

Neutron reflection may also be used for the structural studies of FSLC films. At the boundary between two media with different neutron refractive indices a neutron beam may be totally reflected when incident at angles less than the critical glancing angle.

In the region of this critical angle the neutron reflectivity may be quite sensitive to the inhomogeneities of the refractive-index profile normal to the boundary. The neutron refractive index, in its turn, is simply related to the density and composition of the studied material. It makes the neutron reflection especially useful for the structural investigations of soap films.

The main disadvantage of the neutron in comparison with X-ray reflection, however, is the lack of sensitivity. This arises because of the weak flux and because the contrast in the refractive indices between air and film is less for neutrons than for X-rays.

In practice, in order to increase the critical reflection angle (which is normally less than one degree) and the intensity of the reflected neutron beam, D_2O is added to the solution for the soap film preparation. For dilute ($c < CMC$) soap films, D_2O is mainly absorbed in the film liquid core. Thus measurements of the neutron reflectivity in this case allows one to determine the thickness of this core. These experiments were successfully carried out on some relatively stable, thick (with h of the order of 1000 Å) soap films [89, 90].

REFERENCES

1. Ivanov I.B., Dimitrov D.S., 1988, in *Thin Liquid Films* (ed. I.B. Ivanov), *Surfactant Science Series*, Marcel Dekker, New York, vol. 29, p. 379.
2. Derjaguin B.V., Kussakov M., 1939, *Acta Physicochem. URSS*, **10**, 25.
3. Derjaguin B.V., Titievskaya A.S., 1953, *Kolloidn. Zh.*, **15**, 316.
4. Allan R.S., Charles G.E., Mason S.G., 1961, *J. Colloid Sci.*, **16**, 150.

5. Faetti S., Fronzoni L., Rolla P.A., 1983, *J. Chem. Phys.*, **79**, 1427.
6. Faetti S., Fronzoni L., Rolla P.A., 1983, *J. Chem. Phys.*, **79**, 5054.
7. Meyerhofer D., Sussman A., Williams R., 1972, *J. Appl. Phys.*, **43**, 3685.
8. Scheludko A., 1957, *Kolloid. Z.*, **155**, 39.
9. Mysels K.J., 1964, *J. Phys. Chem.*, **68**, 3441.
10. Exerowa D., Scheludko A., 1971, *C. R. Acad. Bulg. Sci.*, **24**, 47.
11. Kruglyakov P.M., Rovin Yu.G., 1978, *Physical Chemistry of Black Hydrocarbon Films. Biomolecular Lipid Membranes*, Nauka, Moscow (in Russian).
12. Bergeron V., Radke C.J., 1992, *Langmuir*, **8**, 3020.
13. Langevin D., Sonin A.A., 1994, *Adv. Colloid Interface Sci.*, **51**, 1.
14. Matsumoto M., Montandon C., Hartland S., Watanabe A., 1978, *Chem. Eng. Sci.*, **33**, 831.
15. Hartland S., 1967, *Eng. Sci.*, **22**, 1675.
16. Hartland S., 1968, *J. Colloid Sci.*, **26**, 383.
17. Kruglyakov P.M., Ekserowa D.R., 1990, *Foam and Foam Films*, Khimia, Moscow (in Russian).
18. Scheludko A., 1967, *Adv. Colloid Interface Sci.*, **1**, 391.
19. Born E., Wolf E., 1968, *Principles of Optics*, Pergamon Press, Oxford, UK.
20. Popov Yu.N., 1978, *Prib. Tek. Eksp.*, **3**, 237.
21. Kinzer D., 1985, *Mol. Cryst. Liq. Cryst. Lett.*, **1(5)**, 147.
22. Yang K.H., 1988, *J. Appl. Phys.*, **64**, 4780.
23. Chen S.-M., Pan R.-P., Pan C.-L., 1989, *Appl. Opt.*, **28**, 4969.
24. Sonin A.A., Yethiraj A., Bechhoefer J., Frisken B.J., 1995, *Phys. Rev. E*, **52**, 6260.
25. Perez E., Proust J.E., Ter-Minassian-Saraga L., 1988, in *Thin Liquid Films* (ed. I.B. Ivanov), *Surfactant Science Series*, Marcel Dekker, New York, vol. 29, p. 891.
26. Isozaki T., Hiraoka K., Takanishi Y., Takezoe H. *et al.*, 1992, *Liq. Cryst.*, **12**, 59.
27. Morris St.W., Bruyn J.R., May A.D., 1991, *J. Stat. Phys.*, **64**, 1025.
28. Cheung C., Hwang Y.H., Wu X.-I., Choi H.J., 1996, *Phys. Rev. Lett.*, **76**, 2531.
29. Bechhoefer J., Géminard J.-C., Bocquet L., Oswald P., 1997, *Phys. Rev. Lett.*, **79**, 4922.
30. Blinov L.M., 1983, *Electro-Optical and Magneto-Optical Properties of Liquid Crystals*, Wiley, Chichester, UK (Russian version: 1978, Nauka, Moscow).
31. Blinov L.M., Chigrinov V.G., 1994, *Electrooptic Effects in Liquid Crystal Materials*, Springer, New York.
32. Sonin A.A., 1995, *The Surface Physics of Liquid Crystals*, OPA-Gordon and Breach, Amsterdam.
33. Zaininger K.H., Revesz A.G., 1964, *RCA Rev.*, March, 85.
34. Azzam R.M.A., Bashara N.M., 1987, *Ellipsometry and Polarised Light*, Elsevier, Amsterdam.
35. Tompkins H.G., 1993, *A User's Guide to Ellipsometry*, Academic Press, Boston. MA.
36. Bahr Ch., Fliegner D., 1992, *Phys. Rev. A*, **46**, 7657.
37. Bahr Ch., Fliegner D., Booth C.J., Goodby J.W., 1994, *Europhys. Lett.*, **26**, 539.
38. Bahr Ch., Booth C.J., Fliegner D., Goodby J.W., 1996, *Europhys. Lett.*, **34**, 507.
39. Bahr Ch., Booth C.J., Fliegner D., Goodby J.W., 1996, *Phys. Rev. Lett.*, **77**, 1083.
40. Pitchford T., Huang C.C., Pindak R., Goodby J.W., 1986, *Phys. Rev. Lett.*, **57**, 1239.
41. Geer R., Stoebe T., Pitchford T., Huang C.C., 1991, *Rev. Sci. Instrum.*, **62**, 415.
42. Stoebe T., Huang C.C., Goodby J.W., 1992, *Phys. Rev. Lett.*, **68**, 2944.
43. Stoebe T., Huang C.C., 1995, *Int. J. Mod. Phys. B.*, **9**, 2285.
44. Adamson A.W., 1976, *Physical Chemistry of Surfaces*, 3rd edn., Wiley, New York.
45. Stoebe T., Mach P., Huang C.C., 1994, *Phys. Rev. E*, **49**, R3587.
46. Pieranski P., Beliard L., Tournellec J.-Ph., Leoncini X. *et al.*, 1993, *Physica A*, **194**, 364.
47. Stannarius R., Cramer Ch., 1997, *Liq. Cryst.* **23**, 371.
48. Stannarius R., Cramer Ch., 1998, *Europhys. Lett.*, **42**, 43.

49. Miyano K., 1982, *Phys. Rev. A*, **26**, 1820.
50. Brazovskaia M., Dumoulin H., Pieranski P., 1996, *Phys. Rev. Lett.*, **76**, 1655.
51. Rayleigh J.W.S., 1945, *The Theory of Sound*, Dover, New York.
52. Joosten J.G.H., 1988, in *Thin Liquid Films* (ed. I.B. Ivanov), *Surfactant Science Series*, Marcel Dekker, New York, vol. 29, p. 569.
53. Krichevsky O., Stavans J., 1995, *Phys. Rev. Lett.*, **74**, 2752.
54. de Gennes P.G., 1974, *The Physics of Liquid Crystals*, Clarendon Press, Oxford, UK.
55. de Gennes P.G., Prost J., 1993, *The Physics of Liquid Crystals*, 2nd edn., Clarendon Press, Oxford, UK.
56. Kats E.I. and Lebedev V.V., 1994, *Fluctuational Effects in the Dynamics of Liquid Crystals*, Springer, New York (Russian version: 1988, Nauka, Moscow).
57. Chu B., 1991, *Laser Light Scattering*, 2nd edn., Academic Press, New York.
58. Vrij A., 1964, *J. Colloid Sci.*, **19**, 1.
59. Young Ch.Y., Pindak R., Clark N.A., Meyer R.B., 1978, *Phys. Rev. Lett.*, **40**, 773.
60. Rosenblatt Ch., Pindak R., Clark N.A., Meyer R.B., 1979, *Phys. Rev. Lett.*, **42**, 1220.
61. Rosenblatt Ch., Meyer R.B., Pindak R., Clark N.A., 1984, *Phys. Rev. A*, **21**, 140.
62. Dierker S.B., Pindak R., 1987, *Phys. Rev. Lett.*, **59**, 1002.
63. Spector M.S., Sprunt S., Litster J.D., 1993, *Phys. Rev. E*, **47**, 1101.
64. Shalaginov A.N., 1996, *Phys. Rev. E*, **53**, 3623.
65. Sprunt S., Litster J.D., 1987, *Phys. Rev. Lett.*, **59**, 2682.
66. Sprunt S., Spector M.S., Litster J.D., 1992, *Phys. Rev. A*, **45**, 7355.
67. van Winkle D.H., Clark N.A., 1984, *Phys. Rev. Lett.*, **53**, 1157.
68. van Winkle D.H., Clark N.A., 1988, *Phys. Rev. A*, **38**, 1537.
69. Hoffmann E., Stegemeyer H., 1996, *Ferroelectricity*, **179**, 1.
70. Hoffman E., Stegemeyer H., 1996, *Ber. Bunsenges. Phys. Chem.*, **100**, 1250.
71. Becker A., Stegemeyer H., 1997, *Ber. Bunsenges Phys. Chem.*, **101**, 1957.
72. Becker A., Stegemeyer H., 1997, *Liq. Cryst.*, **23**, 463.
73. Brock J.D., Aharony A., Birgeneau R.J., Evans-Lutterodt K.W. *et al.*, 1986, *Phys. Rev. Lett.*, **57**, 98.
74. Brock J.D., Noh D.Y., McClain B.R., Litster J.D. *et al.*, 1989, *Z. Phys. B*, **74**, 197.
75. Clunie J.S., Corkill J.M., Goodman J.F., 1966, *Discuss. Faraday Soc.*, **42**, 34.
76. Sirota E.B., Pershan P.S., Sorensen L.B., Collett J., 1987, *Phys. Rev. A*, **36**, 2890.
77. *Synchrotron Radiation Research: Advances in Surface and Interface Science* (ed. R.Z. Bachrach), Plenum Press, New York, 1992, vols. 1 and 2.
78. Pershan P.S., 1988, *Structure of Liquid Crystal Phases*, World Scientific, Singapore.
79. Bahr Ch., 1994, *Int. J. Mod. Phys. B*, **8**, 3051.
80. Tweet D.J., Holyst R., Swanson B.D., Stragier H., Sorensen L.B., 1990, *Phys. Rev. Lett.*, **65**, 2157.
81. Gierlotka S., Lambooy P., de Jeu W.H., 1990, *Europhys. Lett.*, **12**, 341.
82. Bélorgey O., Benattar J.J., 1991, *Phys. Rev. Lett.*, **66**, 313.
83. Shindler J.D., Mol E.A.L., Shalaginov A., de Jeu W.H., 1995, *Phys. Rev. Lett.*, **74**, 722.
84. Mol E.A.L., Schindler J.D., Shalaginov A.N., de Jeu W.H., 1996, *Phys. Rev. E*, **54**, 536.
85. Cheng M., Ho J.T., Hui S.W., Pindak R., 1987, *Phys. Rev. Lett.*, **59**, 1112.
86. Cheng M., Ho J.T., Hui S.W., Pindak R., 1988, *Phys. Rev. Lett.*, **61**, 550.
87. Chao C.Y., Chou C.F., Ho J.T., Hui S.W. *et al.*, 1996, *Phys. Rev. Lett.*, **77**, 2750.
88. Chou C.F., Jin A.J., Chao C.Y., Hui S.W. *et al.*, 1997, *Phys. Rev. E*, **55**, R6337.
89. Hayter J.B., Highfield R.R., Pullman B.J., Thomas R.K. *et al.*, 1981, *J. Chem. Soc., Faraday Trans. 1*, **77**, 1437.
90. Highfield R.R., Humes R.P., Thomas R.K., Cummins P.G. *et al.*, 1984, *J. Colloid Interface Sci.*, **97**, 367.

4

Thinning, Stratification and Rupture

The issues concerning the thermodynamic stability of free-standing films (e.g. the character of the film thinning and rupture) are very important for consideration of more complex practically applicable systems: emulsions, foams, etc. (e.g. [1]). Here we will describe these issues in more detail. In order to understand better the main physical principles of the film thinning and rupture phenomena, we will firstly analyse them for free-standing amorphous films and only after that will we make some comments on FSLC films. Unfortunately, at the moment, thinning and especially rupture of liquid crystalline films is not so well studied as it is for amorphous ones. This is due to the organised molecular structure of mesophases, making it more difficult to analyse the FSLC film thinning and rupture processes, which are quite complex even for amorphous films.

4.1 MAIN STAGES OF THE THINNING PROCESS

Thinning of a free-standing liquid film is quite a complicated phenomenon, especially when it concerns real agitated systems, such as foams, suspensions and emulsions. However, under certain conditions (like in model experimental cells described in Chapter 3) this process occurs more regularly. For convenience it can be divided into the following stages (see, for example, [2]), Fig. 4.1:

(a) mutual approach along their axis of symmetry of two slightly deformed air bubbles, divided by a fluid;
(b) at a given fluid gap width the curvature in the centre of the system changes its sign and the interfaces acquire a bell-shaped form, called 'dimple';
(c) the dimple height initially increases, then decreases, and, finally, a virtually plane-parallel film forms, which thins further;
(d) the film interface fluctuations, caused by thermal motion or other disturbances, under the action of the disjoining pressure increases their amplitude so much that

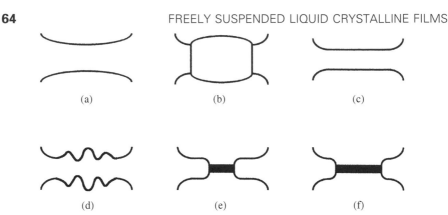

Fig. 4.1 Stages of free-standing film thinning (reprinted from Reference [2] courtesy of Marcel Dekker, Inc.)

under some critical thickness, h_c, the film either ruptures, or a thinner spot of the so-called 'Newton black film' (NBF) forms (see also Chapter 7);
(e) this spot increases its radius and/or coalesces with other spots until the whole film becomes a NBF;
(f) at this moment the equilibrium between the film and its meniscus is violated, and the NBF expands until reaching its equilibrium diameter. The thickness of this equilibrium NBF is usually of the order of several tens of Å.

In the presence of long range repulsive forces, the film can reach equilibrium before h_c. Then the thinning process stops at stage (c), and the so-called 'common black film' (CBF) forms (see also Chapter 7). Its thickness is usually of the order of several hundreds of Å.

Sometimes a plane-parallel film does not occur, but a dimple persists until rupture or the formation of a NBF.

This is a general picture of thinning of any freely suspended liquid film. To analyse more precisely the thinning process, one often needs to take into account the concrete molecular structure of the film and the nature of forces acting in its bulk and at the interfaces. These factors determine, to a large extent, the rate and character of the film thinning.

4.2 MYSELS CLASSIFICATION OF THINNING FREE-STANDING FILMS

According to the composition of surfactant solutions from which they are produced and, as a consequence, to the character of their thinning, soap films have been classified by Mysels *et al.* [3] into three different categories.

1. Rigid or Reynolds—the films with rigid immovable interfaces. Such free-standing films are observed quite rarely. A typical example is the films prepared

from sodium dodecyl sulphate solutions in the presence of dodecanol. The surfactant monolayers which border these films are known to be extremely compact and rigid [4].

2. Simple mobile—films exhibiting a rapid turbulent surface motion. These films are often observed and can be prepared from many dilute ($c <$ CMC) surfactant solutions.

3. Irregular mobile—films consisting of domains with different film thicknesses inside them, separated by walls. The growth and mutual transformation of these domains are observed during thinning of such films. These films are also quite common and are usually prepared from different concentrated ($c >$ CMC) surfactant solutions.

In principle, the above-described general classification may also be applied to all FSLC films. For example, thick thermotropic nematic films can be referred either to the first or to the second categories; thermotropic smectic films and lyotropic films with layered structure fit into the third category.

Films of the first two classes thin gradually by means of transport of a substance (drainage) from the flat portion of the film to the film meniscus. This regime of thinning is also often called 'drainage'.

Films of the third category exhibit a stepwise thinning (stratification). Stratification usually occurs if the film thickness is quite low—about several hundred Å. This is due to the layered ordering of molecules or molecular aggregates inside thin films (see also Chapters 1 and 2). The stratification phenomenon can, in principle, also be observed in a broad film thickness interval, for films prepared from substances with layered structure (e.g. lyotropic lamellar and thermotropic smectic phases).

4.3 GRADUAL THINNING OR 'DRAINAGE'

4.3.1 HYDRODYNAMIC EXPRESSIONS IN LUBRICATION APPROXIMATION

Consider the drainage of a horizontal cylindrical-form amorphous liquid film with plane-parallel interfaces formed between two air bubbles and bounded by a meniscus (Fig. 4.2). During the drainage the thickness h of such a film gradually decreases, and the air bubbles approach one another, i.e. they coalesce.

As has been mentioned already in Chapter 2, the condition $R \gg h$ (where R is the film radius) is almost always fulfilled for free-standing liquid films. This means that the parameter $\delta = h/R \to 0$, i.e. the hydrodynamic equations describing the film drainage can be simplified by neglecting all the terms containing δ. The assumption of δ being small is often called in literature the 'lubrication approximation' (see, for example, [2]).

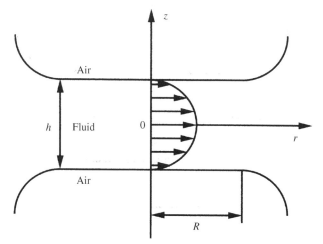

Fig. 4.2 Profile of a thinning fluid film (reproduced from Reference [5] by permission of Academic Press, Inc.)

In the absence of gravity (horizontal thin film) the drainage can be described, in a cylindrical co-ordinate system, by the following hydrodynamic equations, written in the lubrication approximation (e.g. [2, 5, 6]):

$$\frac{\partial P}{\partial r} = \eta \frac{\partial^2 v_r}{\partial z^2}, \tag{4.1}$$

$$\frac{\partial P}{\partial z} = 0, \tag{4.2}$$

$$\nabla v = \frac{1}{r}\frac{\partial(rv_r)}{\partial r} + \frac{\partial v_z}{\partial z} = 0. \tag{4.3}$$

Here η is the viscosity of the liquid forming the film; v_r and v_z are the fluid velocity components in the directions of r and z co-ordinate axes, respectively.

Eq. (4.1) is the Navier–Stokes equation, (4.2) means that the hydrostatic (gravity) effects are neglected, and (4.3) is the medium continuity equation.

The following boundary conditions can also be written:

$$v_r = u \text{ at } z = \pm h/2, \tag{4.4}$$

$$v_z = -V/2 \text{ at } z = \pm h/2 \tag{4.5}$$

$$\frac{\partial v_r}{\partial z} = 0 \text{ at } z = 0, \tag{4.6}$$

Here u is the radial component of the surface velocity, depending on r; $V = -(\mathrm{d}h/\mathrm{d}t)$ is the rate of film thinning.

By integrating the set of equations (4.1)–(4.3) with respect to the boundary conditions (4.4)–(4.6), one easily finds:

$$v_r = \frac{1}{2\eta} \frac{\partial P}{\partial r} \left(z^2 - \frac{h^2}{4} \right) + u(r), \tag{4.7}$$

$$v_z = \frac{1}{2\eta r} \frac{\partial}{\partial r} \left(r \frac{\partial P}{\partial r} \right) \left(\frac{h^2 z}{4} - \frac{z^3}{3} \right) - \frac{z}{r} \frac{\partial (ru)}{\partial r}, \tag{4.8}$$

$$V = \frac{1}{r} \left(2uh - \frac{h^3}{6\eta} \frac{\partial P}{\partial r} \right), \tag{4.9}$$

$$P(r) = P(R) + \Pi(h) - \frac{12\eta}{h^2} \int_r^R u(r)\partial r - \frac{3\eta V}{h^3}(R^2 - r^2), \tag{4.10}$$

where P is the pressure inside the film and Π is the disjoining pressure (see Chapter 2).

4.3.2 REYNOLDS FILMS

Firstly we will analyse the easiest case—the drainage process of a liquid free-standing amorphous film with rigid boundaries, or Reynolds film [7].

Since there is no substance motion at the interfaces of such a film, one should substitute $u = 0$ into Eqs. (4.4) and (4.7)–(4.10). Then, from Eq. (4.10), taking into account (2.25′) with $P_f = P(R)$, we obtain the following radial pressure distribution inside the film:

$$P(r) = P_0 + \frac{3\eta}{h^3} \frac{\partial h}{\partial t}(R^2 - r^2), \tag{4.11}$$

where P_0 is the atmospheric pressure near the surface of the film.

If F is the force acting on the air bubbles, then the following force balance equation can be written:

$$F = \int_0^R 2\pi r[P(r) - P_0]\, \partial r. \tag{4.12}$$

Taking into account (4.11), we have

$$F = -\frac{3\eta \pi R^4}{2h^3} \frac{\partial h}{\partial t}, \tag{4.13}$$

which gives the well-known Reynolds formula for the velocity of thinning of a cylindrical fluid film with rigid boundaries, acted on by an external force F:

$$V = -\frac{\partial h}{\partial t} = \frac{2Fh^3}{3\pi\eta R^4} = V_{\text{Re}}. \tag{4.14}$$

For a liquid film, formed inside a meniscus, Eq. (4.14) can be represented in a more convenient form. Indeed, in this case, $F = \Delta P \pi R^2$, where $\Delta P = P_f - P_m = P_c - \Pi$ is the driving pressure (see Eqs. (2.25') and (2.25'')) [8]

$$V_{Re} = \frac{2\Delta P h^3}{3\eta R^2}.$$ (4.15)

The time dependence of the film thickness $h(t)$ can easily be found by integrating (4.15)

$$h(t) = h_0 \left(1 - \frac{4}{3} h_0^2 \frac{\Delta P}{\eta R^2}\right)^{-1/2}.$$ (4.16)

Here h_0 is the initial film thickness.

The Reynolds formula works quite well for liquid films with really rigid interfaces, for example for films bounded by solid plates. Experimental measurements of the force F acting between two atomically smooth mica plates, separated by an amorphous fluid, as a function of the distance h between the plates, show that Eq. (4.14) is valid for films thicker than 50 nm. For thinner films the stabilising effect of solid boundaries becomes considerable and the film drains more slowly than predicted by Eq. (4.14) [9, 10].

Now some words about thinning of Reynolds FSLC films. Consider, for instance, a thermotropic nematic film. The hydrodynamics of a nematic liquid are similar to those of an isotropic one, with the difference that the viscosity in this case is anisotropic (see, for example, [11, 12]). Thus for small perturbations of the initial (e.g. homeotropic) orientation of the director, the nematic film thinning will also be described by the Eqs. (4.14)–(4.16), but η must be substituted by its value $\eta_\perp$, measured perpendicular to the director orientation [13].

Thinning of thermotropic smectic and of some lyotropic liquid crystalline films will be analysed below in this chapter.

4.3.3 SIMPLE MOBILE FILMS

For free-standing liquid films with mobile surfaces, the Reynolds formula is not very applicable. For these films the drainage velocity is usually several times greater than predicted by Eqs. (4.14) and (4.15).

The mobility of the film interfaces can be taken into account by the boundary condition (4.4). Two cases may be distinguished:

– completely mobile film interfaces, for which $v_r(z = \pm h/2) \neq 0$ and $\partial v_r/\partial z = 0$;
– partially mobile film interfaces, for which $v_r(z = \pm h/2) \neq 0$ and $\partial v_r/\partial z \neq 0$.

By a procedure analogous to that used to obtain the Reynolds formula, it can be shown that the velocity V of thinning of the film with mobile interfaces will

differ from the Reynolds film thinning velocity V_{Re} only by a numerical factor
[14, 15]:

$$V = \frac{4}{n^2} V_{Re}.$$ (4.17)

Here n is the number of immobile film interfaces.

Eq. (4.17) is rigorous only for films with completely mobile or immobile surfaces,
i.e. for integer values of n ($n = 0$, 1, 2). However, it also approximately describes
thinning of films with partially mobile interfaces. In this case n will have non-integer
values. A value of n between 1 and 2 indicates that at least one interface is partially
mobile, and between 0 and 1 that both interfaces are partially mobile [15].

Schematic velocity profiles $v_r(z)$ of draining liquid films with different n are
shown in Fig. 4.3.

For $n = 2$, the v_r profile has a parabolic form (Poiseuille flow). The viscous
dissipation is maximal in this case and $V = V_{Re}$ (Reynolds film).

For $n = 1$, v_r has a half-parabolic form and V is four times higher than for a
Reynolds film.

Finally, for $n = 0$, this profile is of a 'plunge' form, i.e. v_r is constant over the film
thickness and the velocity of thinning is maximal, since the dissipation of energy is
negligible.

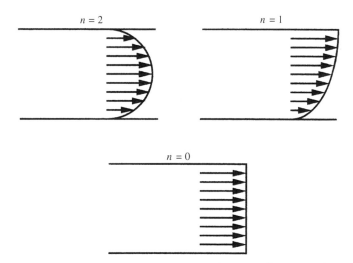

Fig. 4.3 Velocity profiles for a thinning liquid film as a function of mobility of its
interfaces. n is the number of immobile film interfaces (adapted from Reference
[20])

4.3.4 SIMPLE MOBILE SOAP FILMS. ROLE OF SURFACE VISCOELASTICITY

As we already know (see Chapter 2), the surfactant monolayers, which bound the free-standing soap films, may exhibit considerable dilational viscoelasticity. The latter greatly influences the rate of soap film thinning (e.g. [5, 6, 16, 17]) and the degree of stability of foams [18].

In the case of mobile soap films one needs to consider explicitly the balance of horizontal stresses at the film surfaces (i.e. at $z = \pm h/2$) [2]:

$$\eta \frac{\partial v_r}{\partial z} = \eta \left[\frac{6u(r)}{h} - \frac{3rV}{h^2} \right] = \frac{\partial \gamma}{\partial r}. \tag{4.18}$$

The surface tension, γ, is a function of the surfactant concentration at the surface, Γ. This variable is connected with the bulk surfactant concentration, c, by the following equation for mass conservation [2]:

$$\frac{\partial \Gamma}{\partial t} + (\Gamma u) - D_s \Delta_r \Gamma = -D \frac{\partial c}{\partial z}, \tag{4.19}$$

where D_s and D_r, respectively, are the surface and bulk diffusion coefficients, and Δ_r is the radial component of the Laplacian operator.

In Eq. (4.19) the convection term $\mathrm{div}(\Gamma u)$ is roughly equal to $(\Gamma u)/r$. The fractional change of the surface concentration per unit time due to this term is of the order u/r (i.e. typically, $10^{-1}\,\mathrm{s}^{-1}$) [16, 17]. This means that during the drainage process, lasting usually a few tens of seconds, the surface surfactant concentration will decrease substantially. The depleted film regions should be replenished with new surfactant molecules by surface and bulk diffusion. However, the corresponding diffusion rates are negligible, as seen from the numerical comparison between different terms of Eq. (4.19):

$$(\Gamma u) \propto \Gamma u/R \propto 10^{-8},$$

$$D_s \Delta_r \Gamma \propto D_s \Gamma/R^2 \propto 10^{-12},$$

$$D \partial c/\partial z \propto D h \Delta_r c \propto D h c/R^2 \propto 10^{-10}$$

in $\mathrm{g\,cm}^{-2}\,\mathrm{s}^{-1}$, with $D \approx 10^{-5}\,\mathrm{cm}^2/\mathrm{s}$, $D_s \approx 10^{-7}\,\mathrm{cm}^2/\mathrm{s}$, $h \approx 1\,\mu\mathrm{m}$, $R \approx 1\,\mathrm{mm}$, $\Gamma \approx 10^{-4}\,\mathrm{mg/cm}^2$, $c \approx 1\,\mathrm{mg/cm}^3$, and using the lubrication approximation (see above in this chapter) to evaluate $\Delta_r c$ ($\partial^2 c/\partial z^2 = 0$).

So, according to our estimations, we can neglect both bulk and surface diffusions, which in our case are much slower than convection. Eq. (4.19) then becomes

$$\frac{\partial \Gamma}{\partial t} + (\Gamma u) = 0. \tag{4.20}$$

After integrating with respect to time, t, and introducing l as the lateral displacement, we get

$$\frac{\partial \Gamma}{\partial t} + \Gamma \frac{\partial}{\partial r}\left(\frac{1}{r}\frac{\partial rl}{\partial r}\right) = 0. \tag{4.21}$$

Here $l = \int_0^t u(t')\,\partial t'$.

The horizontal force then becomes

$$\frac{\partial \gamma}{\partial r} = \frac{\partial \gamma}{\partial \Gamma}\frac{\partial \Gamma}{\partial r} = \varepsilon\left(\frac{\partial^2 l}{\partial r^2} + \frac{1}{r}\frac{\partial l}{\partial r} - \frac{l}{r^2}\right), \tag{4.22}$$

where we have introduced the surface dilational viscoelasticity, $\varepsilon = S\partial\gamma/\partial S$ (see Eq. 2.7).

For negligible rates of diffusion and in the absence of relaxation processes at the film interfaces $\varepsilon = -\Gamma\partial\gamma/\partial\Gamma = \varepsilon_\partial$ $(\eta_d = 0)$; see Eqs. (2.7) and (2.9). Then the final balance of horizontal stresses is (see Eq. 4.22) [5, 6, 16, 17]

$$\eta\left[\frac{6u(r)}{h} - \frac{3rV}{h^2}\right] = \varepsilon_\partial \int_0^t \left[\frac{\partial^2 u(t')}{\partial r^2} + \frac{1}{r}\frac{\partial u(t')}{\partial r} - \frac{u(t')}{r^2}\right]\partial t'. \tag{4.23}$$

The integration of (4.23) can be performed numerically. Examples of the calculated time dependencies of the film thickness $h(t)$ from Eqs. (4.21) and (4.23) for Reynolds film and mobile soap film with $\varepsilon_d = 34$ dyne/cm, in comparison with the experimental data for the film prepared from water solution of the decylpentaethylene glycol ether $(C_{10}E_5)$ surfactant (CMC $= 0.9$ mmol/l) are shown in Fig. 4.4. It is evident that, indeed, the soap film with mobile surfaces drains much faster than the Reynolds film, i.e. the surface elasticity influences considerably the rate of free-standing soap film thinning.

In the case of lyotropic FSLC films ($c >$ CMC), the contribution of the surface dilational elasticity can often be neglected. The main mechanism of thinning for such films is stratification (see below in this chapter). However, if a dilute ($c <$ CMC) soap film thins at higher thicknesses as a mobile film, and at lower thicknesses stratifies (which is quite often observed experimentally) it is important to take into account ε_d—to characterise the whole process of thinning.

Stratification is also the principle mechanism of thinning for thermotropic FSLC films. However, the surface share viscosity should play a role in this case (see, for example, [11, 12, 19]).

4.3.5 INFLUENCE OF GRAVITY ON THE FILM THINNING

In a real experimental situation, gravity often plays an essential role in the drainage of freely suspended liquid films.

The following factors increase the role of gravity:

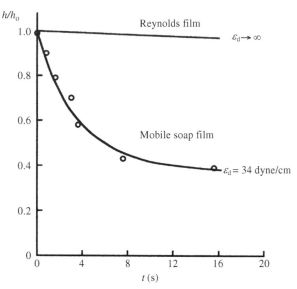

Fig. 4.4 Calculated time dependencies of the thickness h of the Reynolds film and of the mobile soap film with the dilational surface elasticity $\varepsilon_d = 34$ dyne/cm (curves). Experimental data for a $C_{10}E_5$ surfactant film with the surfactant concentration $c = CMC/3$ (points): h_0 is the initial film thickness (reproduced with permission from Reference [16] Copyright 1993 by the American Physical Society)

– considerable film thickness;
– spatial inhomogeneity of the film thickness;
– inclined position of the film with respect to the horizontal plane.

  Gravity inevitably should be taken into account in the analysis of such practically important condensed media as foams, emulsions and suspensions (see, for example, [1, 20–25]). Since these systems consist of multiple, spontaneously oriented, fluid films, all the just-mentioned factors play a role here simultaneously.

  As the simplest example let us consider, following [26], thinning of a two-dimensional amorphous, liquid, Reynolds free-standing film of homogeneous thickness, inclined at an angle α to the horizontal plane (Fig. 4.5). In this case it is convenient to use the Dekarte co-ordinates. The liquid in this film drains along the x co-ordinate under the action of both the applied external force F and the gravity force $F_g = \rho g \sin \alpha$ (ρ is the film fluid density and g is the acceleration due to gravity). It is obvious that the gravity force will increase the resulting driving force, which causes drainage, and consequently accelerates the rate of film thinning.

  Due to the action of gravity, the Navier–Stokes equation (4.1) will be modified as

$$\frac{\partial P}{\partial x} = \eta \frac{\partial^2 v_x}{\partial z^2} + \rho g \sin \alpha. \qquad (4.24)$$

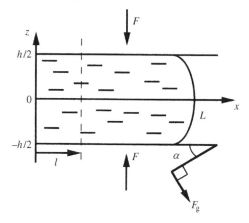

Fig. 4.5 Drainage of a two-dimensional inclined film (reprinted from Reference [26], copyright 1979 with permission from Elsevier Science)

Here v_x is the fluid velocity in the direction of the x axis.

Integration of (4.24) with respect to the boundary conditions $(\partial V_x)/(\partial z) = 0$ at $z = 0$ and $u = 0$ at $z = \pm h/2$ gives the velocity profile as:

$$v_x = \frac{1}{2\eta}\left(\rho g \sin \alpha - \frac{\partial P}{\partial x}\right)\left(\frac{h^2}{4} - z^2\right). \tag{4.25}$$

The volume flow rate, Q, of liquid in the film at a distance x is $2\int_0^{h/2} v_x\,\partial z$ and this must equal the liquid squeezed out of the film over the distance $x - l$, due to each interface of the film approaching the mid-plane $z = 0$ with a rate $v = -\frac{1}{2}(\partial h/\partial t) = V_g/2$ (here V_g is the film thinning velocity in the presence of gravity); see Fig. 4.5. Thus we obtain

$$\frac{Q}{2} = (x - l)v = \frac{h^3}{24\eta}\left(\rho g \sin \alpha - \frac{\partial P}{\partial x}\right). \tag{4.26}$$

Since v and h are not functions of x we may integrate this expression with the boundary condition $P = P_0$ at $x = 0$, which gives

$$P - P_0 = \frac{24\eta v}{h^3}\left(lx - \frac{x^2}{2}\right) + \rho g \sin \alpha x. \tag{4.27}$$

This yields:

$$l = \left(\frac{P_L - P_0}{L} - \rho g \sin \alpha\right)\frac{h^3}{24\eta v} + \frac{L}{2}, \tag{4.28}$$

remembering that $P = P_L$ when $x = L$.

The force acting on the film per unit width is $F = \int_0^L P \partial x$, which gives the following expression for the film thinning velocity:

$$V_g = \frac{F^* h^3}{\eta L^3},\tag{4.29}$$

where $F^* = F - (P_0 + P_L)L/2$.

Expression (4.29) is the generalised two-dimensional analogue of the Reynolds formula (see Eq. 4.14). It can be used, whenever necessary, to express the rate of the film thinning V_g (which is difficult to be measured) by the driving force F^*.

Eq. (4.29) can also be applicable for consideration of FSLC films: for instance, inclined free-standing thermotropic nematic films. In this case η means the corresponding viscosity coefficient of the nematic mesophase.

4.3.6   ROLE OF SURFACE DIMPLES IN THE FILM THINNING

In all our previous analyses we have considered fluid free standing films as having plane-parallel interfaces. However, as we know, in a real experimental situation, it is not always so. Often, films having a bell-form profile or dimple can be formed during the thinning process (see above in this chapter and Fig. 4.1c). The presence of this dimple is due to the pressure gradient inside the film. The thickness of a dimpled film of radius R is maximal in the film centre: $h = h_{max}$ for $r = 0$, and minimal at the film periphery: $h = h_{min}$ for $r = R$ (see Figs. 4.1c and 4.2).

Several attempts to evaluate theoretically the thinning velocity of the dimpled film V_d at the film periphery have been made [27–31]. The obtained expression for V_d differs from the Reynolds formula by a numerical parameter a (see Eqs. (4.14) and (4.17)):

$$V_\partial = -\frac{\partial h_{min}}{\partial t} = \frac{a}{n^2} \frac{\gamma h_{min}^3}{\eta R^2 R_\partial}.\tag{4.30}$$

Here γ is the surface tension and R_d is the dimple radius.

Different authors give different values of a, lying within the limits 1.6–5.6 [27–31]. Nevertheless, taking $n = 2$ (i.e. assuming both film interfaces to be rigid), we obtain from (4.30) that $V_d < V_{Re}$, i.e. that the dimple decreases considerably the rate of thinning. This is in accordance with experimental observations (e.g. [31]).

Note that dimples occur most often in isotropic liquid and dilute soap films, but they may also be observed in freely suspended thermotropic nematic films. The latter circumstance causes no surprise due to the similarity of the viscous properties of nematics and isotropic fluids. It may be interesting, for the nematic films, to study the influence of a dimple on the director distribution.

In the more structurally ordered FSLC films (such as cubic lattice micellar, lyotropic lamellar and thermotropic smectic), however, dimples are not very well pronounced or do not occur at all. This is due to the relatively high bulk viscosity of these films.

More details on dimples can be found, for example, in References [2, 20, 32] (and in the references cited therein).

4.4 STRATIFICATION

4.4.1 STRATIFICATION AND LIQUID CRYSTALLINE ORDERING

Beginning with the first systematic studies by Johnnott and Perrin early in this century (see Section 1.1), the stepwise thinning (or 'stratification') of soap films has been extensively investigated by many authors (e.g. [33–38]). The stratification was observed not only in soap, but also in lyotropic and thermotropic liquid crystalline [6, 39], latex suspension [40], and other structurally ordered films.

These studies showed that stratification is connected with layered (i.e. liquid crystalline, smectic-like) ordering of molecules or molecular aggregates (e.g. micelles) inside the films. These layers flow out of the film surface to the surrounding fluid meniscus, producing in this way a stepwise thinning of the film with a step height that is proportional to the layer thickness l (Fig. 4.6).

It is evident that lyotropic lamellar or thermotropic smectic films, which consist of molecular monolayers or bilayers, can, in principle, stratify. However, the cases of other stratifying FSLC films (for example, lyotropic micellar or thermotropic nematic) demand additional comments.

To explain the stratification of soap films consisting of spherical micelles the following model, which is in agreement with experimental observations, has been proposed in References [33, 34]. It postulates that a soap film consists of effective spherical micelles of diameter $\delta \approx 100$ Å. Each effective micelle, in its turn, represents a surfactant micelle, itself surrounded by water molecules. When the thickness, h, of the soap film is quite high (experimentally, $h \gg 1000$ Å, i.e. $h \gg \delta$), the effective micelles inside it are disordered (Fig. 4.7a). Such a film cannot stratify and thins gradually. When h becomes lower (experimentally, $h < 1000$ Å, i.e. $h \approx \delta$), the intermolecular interactions are growing and the

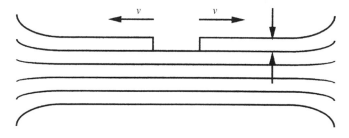

Fig. 4.6 Schematic representation of free-standing liquid film stratification

effective micelles organise themselves in a cubic lattice with the mean distance δ between their gravity centres (Fig. 4.7b). Such a film will stratify by means of the displacement of the obtained micellar layers from the flat portion of the film to the plateau borders. The stratification step in this case, naturally, equals the integer number of the thicknesses δ.

Now some words concerning the stratification of thermotropic nematic free-standing films. The existence of smectic layers near the surfaces of nematics was predicted theoretically [41–45] and observed experimentally by means of the Israelachvili technique for force measurements (for nematic films bounded by solid surfaces) [46], and by means of synchrotron X-ray scattering (for free interfaces of some nematic materials) [47, 48]. Generally, these smectic layers orient themselves parallel to the nematic surface, stabilising in this way the homeotropic orientation of the director [11, 12, 41–48]. The degree of smectic ordering decays exponen-

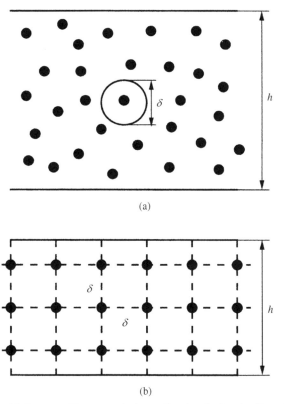

(a)

(b)

Fig. 4.7 (a) A thick soap film consisting of spherical micelles: $h \gg \delta$ (δ is the effective micelle diameter). (b) A thin soap film with cubic lattice ordering of spherical micelles: $h \approx \delta$ (reproduced by permission of Academic Press from Reference [33])

tially from the surface of a nematic with a characteristic decay length equal to the longitudinal smectic correlation length, $\xi_{\parallel}$ [47, 48]. Far from the nematic–smectic A phase transition temperature, T_{NA}, $\xi_{\parallel} \approx 100\,\text{Å}$. Near T_{NA} $\xi_{\parallel}$ diverges: $\xi_{\parallel} \propto (T - T_{NA})^{\nu_{\parallel}}$, where the experimental values of the critical exponent $\nu_{\parallel}$ vary between 0.5 and 0.75 for different materials [12]. One expects considerable surface smectic ordering in nematic substances possessing large longitudinal dipole moments (e.g. cyanobiphenyl derivatives) and in nematics that have a transition to some smectic phase at low temperatures.

Thus if the nematic free-standing film thickness, h, is much greater than $\xi_{\parallel}$, the bulk portion of the film is in a nematic phase, and hence the film may gradually thin (Fig. 4.8a). When $h \approx \xi_{\parallel}$, the whole film should have a smectic-like structure and hence can stratify (e.g. [39, 49–52]), Fig. 4.8(b). Fig. 4.9 illustrates the stratification phenomenon in freely suspended nematic 4′-n-octyl-4-cyanobiphenyl (8CB) film in the vicinity of the nematic–smectic A phase transition. This is the dependence of the film thickness, h, upon the number of stratification steps, N, obtained from the light scattering data.

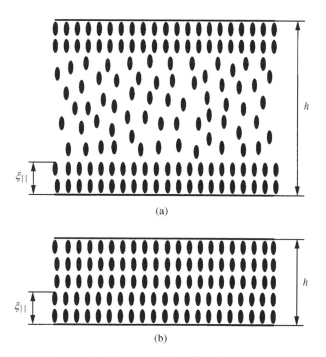

Fig. 4.8 (a) A thick nematic free-standing film: $h \gg \xi_{\parallel}$ ($\xi_{\parallel}$ is the longitudinal smectic correlation length). (b) A thin nematic film with smectic-like ordering: $h \approx \xi_{\parallel}$

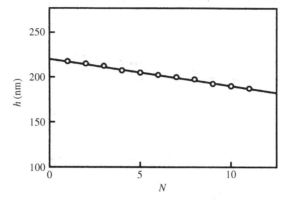

Fig. 4.9 The experimental dependence of the thickness of a draining free-standing film of 8CB liquid crystal at $T - T_{NA} = 0.54\,K$ (T_{NA} is the temperature of the nematic–smectic A phase transition) upon the number of stratification steps (N), which shows discreet jumps with a height of 3 nm (reproduced from Reference [52] by permission of EDP Sciences)

4.4.2   STRATIFICATION DOMAINS AND THEIR DYNAMICS

Even in the first experimental studies of Johnnott and Perrin (see Section 1.1), it has been noted that in the course of the stratification process, a soap film is not homogeneous in its thickness, h. Several areas (up to 8 or 9 [40]) with different h can coexist simultaneously in the film. Seen under a microscope in reflected light, the thinner of these areas look darker than the thicker ones. The just-mentioned areas, which are usually called the 'stratification domains', can be of various geometrical forms. For instance, in the case of circular-shaped free-standing films prepared from thermotropic nematics or from solutions of spherical micelles, stratification domains are also of circular form [Fig. 4.10(a–c)]; for circular films, made from swollen lamellar phase, these domains possess quite complicated borders [like the contours of continents at a geographical map; Fig. 4.10(d)].

During the film thinning the excess liquid flows out from the stratification domains to the surrounding thicker portion of the film and to the Plateau borders. Thus the stratification domains increase in their area, i.e. surface hydrodynamics is observed.

As an example, consider some experimental results on the dynamics of expansion of the stratification domains over the surface of a freely suspended circular-shaped soap film, consisting of spherical micelles (a film prepared from the dodecyltrimethylammonium bromide (DTAB) surfactant water solution with $c \approx 7\,CMC$, $CMC = 15\,mmol/l$). At the beginning of the thinning process stratification domains with obscure borders appear at the surface of such a film (domains V). Later on, the thinner domains (IV and III) with sharp circular borders start to grow, Fig. 4.10(b). And finally, the thinnest domains (II and I) bounded by circular borders with liquid

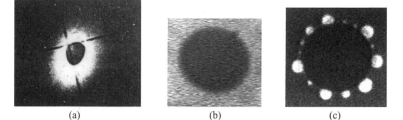

(a)　　　　　　　(b)　　　　　　　(c)

(d)

Fig. 4.10 Examples of different stratification domains observed in circular-shaped freely suspended films, prepared in a Scheludko cell (viewed in the reflecting microscope): (a) circular domains in a thermotropic nematic 5CB film [39] (courtesy of Marcel Dekker); (b) thicker circular domains in a soap film (a water solution of the surfactant dodecyltrimethylammonium bromide, DTAB, with $c \approx 7$ CMC), which consists of spherical micelles (A.A. Sonin, unpublished work); (c) thinner domains in the same film [37] (by permission of EDP Sciences); (d) complicated contours of the stratification domains in a swollen lamellar phase film (prepared from 1.1 wt.% water solution of N,N-dimethyl octadecylamine oxide surfactant) (reproduced from reference [6] by permission of Elsevier Science)

flowing out in the form of small round drops [rim instability; Fig. 4.10(c)], can be observed (e.g. [6, 35–37]).

The rim instability is characteristic of liquids flowing over highly curved surfaces and is due to the counterbalance between the gravity and the surface tension forces. It was first observed by Rayleigh for liquid jets flowing down under the action of gravity [53]. Other examples of such instability are liquid drops placed on thin hair or thread, a spider's web, etc.

The typical time dependence of the thickness h in the central portion of a micellar soap film, measured in the Scheludko cell by the laser interferrometric method (see

Chapter 3), is represented in Fig. 4.11. The area of the $h(t)$ curve corresponding to the thickest stratification domain V is not visible at this dependence, while the areas showing the thinner domains IV, III, II and I, passing through the laser beam, are quite pronounced. Their thicknesses are, respectively, 70, 50, 25 and 8 nm. Small peaks at the $h(t)$ dependence between the areas III–II and II–I show the increase in the intensity of the laser light reflected from the passing border liquid drops [see Fig. 4.10(c)].

An interesting question is how the diameter D of stratification domains changes with time t? Experimental data obtained from the analysis of the recorded video images of stratifying soap films show that for the thickest domains V–III, $D \propto t^2$, while for the thinnest domains II and I (with rim instabilities), $D \propto t$ [5, 36, 37].

It is easy to obtain quantitative relations between D and t assuming that the soap film consists of spherical effective micelles, organised in a cubic lattice [see Fig. 4.7(b)]. Indeed, in this case, the following expression for the dynamics of effective micelles can be written [37]:

$$\frac{\partial S}{\partial t} = J\delta^2 + L\rho i \delta^2, \qquad (4.31)$$

where $S = \pi D^2/4$ and $L = \pi D$ are, respectively, the area and the length of the border of the stratification domain; δ^2 is the area occupied by one effective micelle at the film surface; J is the total number of these micelles flowing through the stable

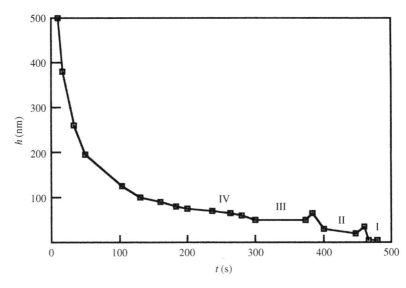

Fig. 4.11 Typical time dependence of the film thickness, h, measured in the central part of a soap film (a water solution of DTAB surfactant with $c \approx 7$ CMC), consisting of spherical micelles. The $h(t)$ curve areas IV, III, II and I show corresponding stratification domains passing through a laser beam (reproduced by permission of EDP Sciences from Reference [37])

stratification domain periphery per unit time, i.e. the outflux from the stable domain border; ρ is the linear density of the rim instability drops: $\rho = N/L$ (N is the total number of these drops at the border of a stratification domain); i is the number of effective micelles flowing through each such drop per unit time, i.e. the outflux from each drop [see Fig. 4.10(c)].

The first term in Eq. (4.31) describes the liquid outflux through the stable periphery of the stratification domain. It means that the physical conditions at the soap film periphery remain unchanged, i.e. the chemical potentials in the film and in the fluid meniscus are the same ($J = $ const). The second term reflects the liquid outflux through the rim instability drops. According to the experimental data of [37], ρ in this term is constant (i.e. $N \propto D$), and the mean value of i is supposed to be the same for each rim instability drop.

Eq. (4.31) can be applied under the following conditions:

(i) $D \ll D_f$ (where D_f is the film diameter), i.e. when the influence of the film border is negligible;

(ii) $D \gg D_0$ (where D_0 is the experimentally observable nucleation diameter of the stratification domain: usually, $D_0 \sim$ several μm [36]). Indeed, the complicated character of the $D(t)$ dependence in the vicinity of D_0 has been experimentally observed in Ref. [35]. This is due to the fluctuations of D. This means that, in the framework of our approach, we cannot determine precisely the integration constant for Eq. (4.31): $D(t = 0) = $ const. But, in any case, this constant is rather small (of the order of magnitude of D_0) and can be neglected.

For the stratification domains with stable borders (domains V–III) the second term in (4.31) vanishes, and we obtain [35]

$$D = 2\delta(Jt/\pi)^{1/2}. \qquad (4.32)$$

Eq. (4.32) allows us to determine the value of the outflux J from the experimental data on $D(t)$. For instance, for the above-considered DTAB surfactant film (with $c \approx 7$ CMC) $J = 10^4 \, \mathrm{s}^{-1}$ [37].

It is most probable that the rim instability liquid drops (which exhibit intensive hydrodynamics inside them) have rather a disordered structure, i.e. the viscosity there is lower than for the cubically ordered surrounding film. This means that, in the presence of rim instability, the liquid flows out from the stratification domain mainly from these drops: i.e. $i \gg J$. Then, for a domain with rim instability, the first term in Eq. (4.31) can be neglected, and we find that D is linearly proportional to t [37]:

$$D = 2\rho i\delta^2 t. \qquad (4.33)$$

Thus the stratification domains with rim instability grow with constant velocity v:

$$v = 2\rho i\delta^2. \qquad (4.34)$$

Experimental values of v for the same DTAB surfactant soap film are: 16 μm/s (domain II) and 97 μm/s (domain I). The corresponding values of the outflux, i, determined from Eq. (4.34), are $7 \times 10^6\,\mathrm{s}^{-1}$ and $4 \times 10^7\,\mathrm{s}^{-1}$ [37], i.e. indeed, $i \gg J$.

4.4.3 ANALOGY BETWEEN STRATIFICATION, WETTING AND DEWETTING PHENOMENA

Wetting and dewetting of solids by fluids are formally identical to the above-described soap film stratification (see, for example, [6, 37]). Indeed, the length L of a thin precursor film moving ahead of the liquid drop wedge, spreading over a solid surface [Fig. 4.12(a)], is proportional to the square root of time (as are the diameters of the stratification domains with stable borders) [54–57]:

$$L \propto (D_{\mathrm{eff}} t)^{1/2}. \tag{4.35}$$

Here $D_{\mathrm{eff}} = -h^3(L)/3\eta(\partial\Pi/\partial h)_{h=h(L)}$ is the effective diffusion coefficient, $h(L)$ the minimal stable thickness of the precursor film, $h(x)$ is its local thickness, η is the liquid viscosity, and Π is the disjoining pressure.

For example, in the case of the dominant van der Waals interaction: $D_{\mathrm{eff}} = A/[6\pi\eta h(L)]$, where A is the Hamaker constant (see Eq. 2.15). Taking the

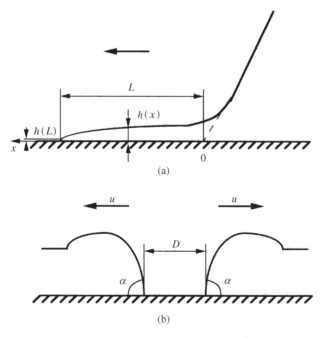

Fig. 4.12 Dynamics of liquid drops on a solid surface: (a) wetting; (b) dewetting

typical values $A \approx 10^{-13}$ erg, $\eta \approx 1\,\mathrm{Ps}\,\mathrm{e}$, $h(L) \approx 10^{-6}\text{--}10^{-5}$ cm, we get $D_{\mathrm{eff}} \approx 10^{-9}\text{--}10^{-8}$ cm²/s. The obtained estimation of D_{eff} is comparable to the coefficient $4\delta^2 J/\pi \propto 10^{-8}$ cm²/s, determined for the stratification domains with stable borders (see Eq. 4.32).

In the case of a fluid film dewetting a solid surface, the liquid rim (or rim instability), similar to those observed for thinner stratification domains in soap films, is often formed at the border of the dewetted spot of a solid (Fig. 4.12(b)). The diameter D of the spot increases linearly with time, as does the diameter of a stratification domain with rim instabilities, i.e. this spot also grows with a constant velocity u [54, 58–60]:

$$u = 2^{-3/2} \left(\frac{\gamma}{\eta} \right) \alpha^3. \tag{4.36}$$

Here γ is the surface tension of the liquid and α is the static contact angle.

Formula (4.36) also describes quite well the expansion dynamics of the stratification domains with rim instabilities. Indeed, for typical soap films: $\gamma/\eta \approx 3 \times 10^3$ cm/s and $\alpha \approx 1^\circ$ [6]. Thus from Eq. (4.36) we obtain $u \approx 50\,\mu\mathrm{m/s}$, which is consistent with the expansion velocities v for these stratification domains (see previous section).

4.5 RUPTURE OF FREE-STANDING FILMS

As we already know (see earlier in this chapter), at the final stages of thinning, two situations may occur:

– the liquid film reaches its equilibrium state—the black film;
– at some critical thickness (usually of the order of several hundreds of Å) the film becomes unstable and ruptures.

The structure and some properties of black films will be described in Chapter 7. Here we will focus our attention on consideration of the rupture process.

Consider firstly a quite simple model of the spontaneous fluctuational rupture of an amorphous liquid free-standing film, first proposed by Scheludko (e.g. [61]) and elaborated further on by de Vrij [62]. As we already know, surface thermal fluctuations are always present at the free-standing film interfaces (see Chapter 3, Figs. 3.11 and 4.1d). Under certain conditions some of these fluctuations may grow until they break the film. It has been shown formally that the mechanism of the film rupture is analogous to the phase separation by spinodal decomposition in solutions [62, 63].

The mean-square amplitude of the film thickness fluctuations (i.e. fluctuations in a squeezing mode, Fig. 3.10) is found to be [62]:

$$\langle h_q^2 \rangle = \frac{2k_{\mathrm{B}}T}{S[\gamma q^2 - 2\partial\Pi/\partial h)]}, \tag{4.37}$$

where k_B is the Boltzmann constant, T is the absolute temperature, S is the film area, γ is the surface tension, q is the wave vector of the fluctuation and h_q is the Fourier component of h: $h(r) = \sum_q h_q e^{iqr}$, r being the co-ordinate of the surface plane.

It is clear from Eq. (4.37) that when $\Pi' = \partial\Pi/\partial h$ is positive there is a wave vector $q^* = \sqrt{2\Pi'/\gamma}$, for which the amplitude of the fluctuations diverges. Π' has positive values at film thicknesses where the electrostatic repulsion is overcome by the van der Waals attraction (see Fig. 2.10).

When the time evolution of the film thickness fluctuations is calculated, it is found that all fluctuations with the wave vectors $q < q^*$ increase exponentially with time:

$$h_q = h_q(0) \exp(t/\tau_q). \tag{4.38}$$

The increase rate can be evaluated for each Fourier component assuming that the film interfaces are rigid (the Reynolds film). One thus finds

$$\frac{1}{\tau_q} = \frac{h^3}{12\eta}(\Pi'q^2 - \gamma q^4/2). \tag{4.39}$$

The faster rate relates to $q_m = q^*/\sqrt{2}$, which gives the characteristic time corresponding to the maximal velocity of the evolution of fluctuations:

$$\tau_m = 24\gamma\eta/[h^3(\Pi')^2]. \tag{4.40}$$

For overcoming van der Waals interaction in the film, one obtains from Eqs. (4.40) and (2.15)

$$\tau_m = 96\pi^2\gamma\eta h^5 A^{-2}, \tag{4.41}$$

i.e. τ_m varies very rapidly with the film thickness.

The breaking time of the film τ_b is estimated as the time for which $\langle h_q^2 \rangle \approx h^2$ and is found to be comparable to τ_m:

$$\tau_b = \tau_m f(h, \gamma). \tag{4.42}$$

Here f ranges from 4.5 to 6.9 for $h = 100$–1000 Å, using $\gamma = 30$ dynes/cm.

The lifetime τ_l of the film is the sum of the thinning (τ_t) and of the breaking (τ_b) times:

$$\tau_l = \tau_t + \tau_b. \tag{4.43}$$

In contrast to τ_b, τ_t decreases when h increases. Indeed, for prevailing van der Waals interactions, assuming that $\Delta P = \Pi_{vdW}$ one obtains from Eqs. (4.14) and (2.15):

$$\partial h/\partial t = -A/9\pi\eta R^2. \tag{4.44}$$

The optimal wave vector leading to film rupture is therefore the one for which $\tau_t + \tau_b$ is minimal, i.e.

$$\partial\tau_t/\partial h + \partial\tau_b/\partial h = 0. \tag{4.45}$$

Furthermore, $d\tau_b/dh$ may be obtained from Eqs. (4.41) and (4.42). Neglecting the dependence of f on h and using (4.44) and (4.45) one obtains the critical thickness h_c, at which the film rupture:

$$h_c = 0.222(AR^2/f\gamma)^{1/4}.$$ (4.46)

Note that formula (4.46) can also be generalised for the case where both van der Waals and electrostatic components of the disjoining pressure are present (see, for example, [18]).

Eq. (4.46) leads to an h_c of the order of 100 Å for films of radii $R = 0.1$ mm, which is in good qualitative agreement with some experiments [64].

However, discrepancies are found in many circumstances. The critical thickness does not vary with the film radius as predicted by the described above model [64]. This can be attributed to dimple formation; indeed, when this is the case, rupture always occurs in the thinnest part of the film, i.e. at the film periphery (see above in this chapter). The value of h_c also does not vary with surface tension as predicted by Eq. (4.46) [6].

The described mechanism of free-standing film rupture is not universal. In the case of low film thicknesses ($h < 100$ Å), hole nucleation in a film may be energetically favourable (e.g. [1, 65]). This hole starts to grow and its expansion leads to film rupture. Formation of such holes has been observed, for example, in dilute soap films [65], in emulsion films submitted to the action of a high voltage [66] and in soap films burst by an electric spark [67] in different amphiphile bilayers (black films) bombarded by α-particles [68].

Being formed in a soap film, such a hole expands with a velocity of the order of magnitude of 100 m/s [67], i.e. much faster than the stratification domains (see above in this chapter). The experimental observation of such rapid processes is very difficult and demands special equipment: for instance, high-speed photo and video cameras [66, 67, 69], reflection from different film portions of two laser beams [67], etc.

Considering the balance of forces acting at a hole edge propagating in a free-standing film, it is easy to show that a hole should expand with a maximal constant velocity v_h (see, for example, [70]):

$$v_h = \left(\frac{2\gamma}{h\rho}\right)^{1/2}.$$ (4.47)

Here γ and ρ are, respectively, the surface tension and the density of the film fluid; h is the film thickness.

Eq. (4.47) is a more precise version of the formula obtained in earlier works by Dupré and Rayleigh (see paragraph 1.1), where v_h was exaggerated by a factor of 2.

Now some comments concerning rupture of FSLC films. Since the hole propagation process has not been specially studied for FSLC films (only thermodynamically stable pores in free-standing ferroelectric smectic films, appearing under the action of an electric field and not leading to the film rupture, have been analysed theoretically [71]), we will not consider this issue here.

The rupture of nematic films can be characterised well by the above-described fluctuational model, due to the affinity of hydrodynamical properties of nematics and isotropic fluids. For thin ($h < 0.1$ μm) nematic films the main components in the disjoining pressure, which should be taken into account in Eq. (4.37), are the electrostatic, Π_e, and the van der Waals, Π_{vdW}. In the case of quite thick nematic films ($h \approx 1$–0.1 μm), however, one should necessarily take into account the elastic disjoining pressure, Π_{el} (see Chapter 2), which is dominant, in this interval of h, over Π_e and Π_{vdW}. The term Π_{el}, as we know, is due to the repulsive elastic forces and hence stabilises the film.

What concerns the layered-structure films, the most probable rupture mechanism, which in this case is the layers fusion initiated by the surfactant concentration fluctuations. This kind of rupture was experimentally observed in thin films, consisting of surfactant bilayers, and in lipid membranes (see, for example, [72]). A schematic representation of such a rupture process is given in Fig. 4.13. At the first stage two surfactant bilayers approach one another in a place where the surfactant concentration fluctuation, i.e. a local lowering of c, occurs [Fig. 4.13(a)]. Then, two approaching bilayers fuse locally in the internal part of the film [Fig. 4.13(b)]. This stage of the film fusion is often called 'hemifusion'. Later on, two external monolayers slide out from the fused portion of the film, leaving only two internal monolayers, fused into one bilayer [Fig. 4.13(c)]. Finally, the total removal of the central bilayer, i.e. the film rupture, takes place [Fig. 4.13(d)].

We do not know of experimental evidence for the realisation of the just-described mechanism of rupture in thermotropic smectic films. However, the fusion of

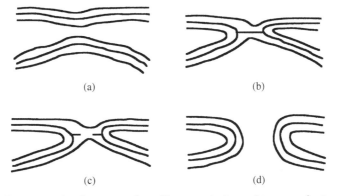

(a)   (b)

(c)   (d)

Fig. 4.13 Rupture of a free-standing film consisting of two surfactant bilayers by means of fusion: (a) local mutual approach of bilayers; (b) their hemifusion; (c) sliding out of two outer monolayers; (d) total removal of the internal fused bilayer, i.e. film rupture (reprinted with permission from Reference [72]. Copyright 1989 American Association for the Advancement of Science)

molecular layers in smectics with the formation of edge dislocations is observed quite often (see, for example, [11, 12]).

REFERENCES

1. Kruglyakov P.M., Ekserowa D.R., 1990, *Foam and Foam Films*, Moscow, Khimia (in Russian).
2. Ivanov I.B., Dimitrov D.S., 1988, in *Thin Liquid Films* (ed. I.B. Ivanov), *Surfactant Science Series*, Marcel Dekker, New York, vol. 29, p. 379.
3. Mysels K.J., Shinoda K., Frankel S., 1959, *Soap Films*, Pergamon Press, New York.
4. Shah D.O., Djabbarah N.F., Wasan D.T., 1978, *Colloid Polym. Sci.*, **256**, 1002.
5. Tambe D.E., Sharma M.M., 1991, *J. Colloid Interface Sci.*, **147**, 137.
6. Langevin D., Sonin A.A., 1994, *Adv. Colloid Interface Sci.*, **51**, 1.
7. Reynolds O., 1886, *Philos. Trans. R. Soc. London*, **177**, 157.
8. Scheludko A., 1967, *Adv. Colloid Interface Sci.*, **1**, 391.
9. Chan D.Y.C., Horn R.G., 1985, *J. Chem. Phys.*, **83**, 5311.
10. Israelachvili J.N., 1986, *J. Colloid Interface Sci.*, **110**, 263.
11. de Gennes P.G., 1974, *The Physics of Liquid Crystals*, Clarendon Press, Oxford, UK.
12. de Gennes P.G., Prost J., 1993, *The Physics of Liquid Crystals*, 2nd edn., Clarendon Press, Oxford, UK.
13. Dimitrov D.S., Ivanov I.B., 1975, *C. R. Acad. Bulg. Sci.*, **28**, 1513.
14. Hartland S., 1967, *Chem. Eng. Sci.*, **22**, 1675.
15. Hartland S., Wood S.M., 1973, *AIChE J.*, **19**, 810.
16. Sonin A.A., Bonfillon A., Langevin D., 1993, *Phys. Rev. Lett.*, **71**, 2342.
17. Sonin A.A., Bonfillon A., Langevin D., 1994, *J. Colloid Interface Sci.*, **162**, 323.
18. Wantke K., Malysa K., Lunkenheimer K., 1994, *Colloids Surf. A*, **82**, 183.
19. Pindak R., Bishop D.J., Sprenger W.O., 1980, *Phys. Rev. Lett.*, **44**, 1461.
20. Palermo T., 1991, *Rev. Inst. Fr. Pet.*, **46**, 325.
21. *Foams*, 1976 (ed. R.J. Akers), Academic Press, London.
22. Hartland S., 1988, in *Thin Liquid Films* (ed. I.B. Ivanov), *Surfactant Science Series*, Marcel Dekker, New York, vol. 29, p. 663.
23. Kruglyakov P.M., 1988, in *Thin Liquid Films* (ed. I.B. Ivanov), *Surfactant Science Series*, Marcel Dekker, New York, vol. 29, p. 767.
24. Malhotra A.K., Wasan D.T., 1988, in *Thin Liquid Films* (ed. I.B. Ivanov), *Surfactant Science Series*, Marcel Dekker, New York, vol. 29, p. 829.
25. Rosen M.J., 1989, *Surfactants and Interfacial Phenomena*, 2nd edn., Wiley, New York.
26. Hartland S., 1979, *Chem. Eng. Sci.*, **34**, 485.
27. Frankel S.P., Mysels K.J., 1962, *J. Phys. Chem.*, **66**, 190.
28. Princen H.M., 1963, *J. Colloid Sci.*, **18**, 178.
29. Buevich Yu.A., Lipkina E.Kh., 1975, *Zh. Prikl. Mekh. Tekh. Fiz.*, **2**, 80.
30. Buevich Yu.A., Lipkina E.Kh., 1978, *Kolloidn. Zh.*, **40**, 167.
31. Chen J.D., Hahn P.S., Slattery J.C., 1984, *AIChE J.*, **30**, 622.
32. Joye J.-L., Miller C.A., 1992, *Langmuir*, **8**, 3083.
33. Nikolov A.D., Kralchevsky P.A., Ivanov I.B., Wasan D.T., 1989, *J. Colloid Interface Sci.*, **133**, 13.
34. Kralchevsky P.A., Nikolov A.D., Wasan D.T., Ivanov I.B., 1990, *Langmuir*, **6**, 1180.
35. Bergeron V., Jiménez-Laguna A.I., Radke C.J., 1992, *Langmuir*, **8**, 3027.
36. Langevin D., Sonin A.A., 1993, *Prog. Colloid Polym. Sci.*, **93**, 357.
37. Sonin A.A., Langevin D., 1993, *Europhys. Lett.*, **22**, 271.

38. Krichevsky O., Stavans J., 1995, *Phys. Rev. Lett.*, **74**, 2752.
39. Perez E., Proust J.E., Ter-Minassian-Saraga L., 1988, in *Thin Liquid Films* (ed. I.B. Ivanov), *Surfactant Science Series*, Marcel Dekker, New York, vol. 29, p. 891.
40. Nikolov A.D., Wasan D.T., Kralchevsky P.A., Ivanov I.B., 1988, in *Ordering and Organisation in Ionic Solutions*, World Scientific, Singapore, p. 302.
41. Rosenblatt Ch., Ronis D., 1981, *Phys. Rev. A*, **23**, 305.
42. Kanel H.V., Litster J.D., Melngailis J., Smith H.I., 1981, *Phys. Rev. A*, **24**, 2713.
43. Rosenblatt Ch., 1984, *Phys. Rev. Lett.*, **53**, 791.
44. Sonin A.A., 1988, *Kristallografiya*, **33**, 697.
45. Poniewierski A., Samborski A., 1995, *Phys. Rev. E*, **51**, 4574.
46. Horn R.G., Israelachvili J., Perez E., 1981, *J. Phys. (France)*, **42**, 39.
47. Als-Nielsen J., Christensen F., Pershan P.S., 1982, *Phys. Rev. Lett.*, **48**, 1107.
48. Pershan P.S., Als-Nielsen J., 1984, *Phys. Rev. Lett.*, **52**, 759.
49. Proust J.E., Perez E.E., 1977, *J. Phys. (France) Lett.*, **38**, 91.
50. Manev E., Proust J.E., Ter-Minassian-Saraga L., 1977, *Colloid Polym. Sci.*, **255**, 1133.
51. Rosenblatt Ch., Amer N.M., 1980, *Appl. Phys. Lett.*, **36**, 432.
52. Böttger A., Joosten J.G.H., 1987, *Europhys. Lett.*, **4**, 1297.
53. Rayleigh J.W.S., 1892, *Philos. Mag.*, **34**, 145.
54. Leger L., Joanny J.F., 1992, *Rep. Prog. Phys.*, 431.
55. Joanny J.F., de Gennes P.G., 1986, *J. Phys. (France)*, **47**, 121.
56. de Gennes P.G., Cazabat A.M., 1990, *C. R. Acad. Sci.*, Ser. II, **310**, 1601.
57. Fraysee N., Valignat M.P., Cazabat A.M., Heslot F., Levinson P., 1993, *J. Colloid Interface Sci.*, **158**, 27.
58. de Gennes P.G., 1987, in *Physics of Amphiphilic Layers* (eds. J. Meunier, D. Langevin, N. Boccara), Springer, Berlin, vol. 34, p. 64.
59. Redon C., Brochard-Wyart F., Rondelez F., 1991, *Phys. Rev. Lett.*, **66**, 715.
60. Brochard-Wyart F., Redon C., 1992, *Langmuir*, **8**, 2324.
61. Scheludko A., 1984, *Colloidal Chemistry*, Mir, Moscow (in Russian).
62. de Vrij A., 1966, *Discuss. Faraday Soc.*, **42**, 23.
63. Cahn, 1965, *J. Chem. Phys.*, **42**, 93.
64. Sonntag H., Strenge K., 1969, *Coagulation and Stability of Disperse Systems*, Halsted-Wiley, New York.
65. de Vries A., 1960, *Proc. Third Int. Congress on Detergency*, Cologne, vol. 2, p. 566.
66. Charles G.E., Mason S.G., 1960, *J. Colloid Sci.*, **15**, 236.
67. Mysels K.J., Vijayendran B.R., 1973, *J. Phys. Chem.*, **77**, 1692.
68. Exerowa D., Kashchiev D., Platikanov D., 1992, *Adv. Colloid Interface Sci.*, **40**, 201.
69. Liang N.Y., Chan C.K., Choi H.J., 1996, *Phys. Rev. E*, **54**, R3117.
70. Culick F.E.C., 1960, *J. Appl. Phys.*, **31**, 1128.
71. Prost J., Lejcek L., 1989, *Phys. Rev. A*, **40**, 2672.
72. Helm C.A., Israelachvili J.N., McGuiggan P.M., 1989, *Science*, **246**, 919.

5

Orientational and Phase Transitions

Talking in this chapter about orientational transitions, we will have in mind phenomena occurring mostly in thermotropic nematic free-standing films, since, in such films, the tilt in the director orientation does not induce phase transfers. Indeed, in smectic films for instance, the tilt of the director often changes the phase state: in smectic A films the director is perpendicular to the smectic layers, while in smectic C films it is inclined relative to the layers (see Fig. 1.2). Only one example of orientation transfer in ferroelectric smectic C* free-standing films will be briefly described.

Considering, further on, phase transitions in FSLC films, we will deal mostly with different thermotropic smectic films, for which a sufficient number of experimental data are available.

5.1 ORIENTATIONAL TRANSITIONS IN NEMATIC FILMS

Orientational transitions in nematic freely suspended films, i.e. the changes of the director orientation, can be divided into the two following groups:

1. spontaneous transitions, i.e. taking place without application of an external force. Such orientational transfers may be initiated, for instance, by variations of the film thickness or the temperature.
2. force-induced transitions, i.e. taking place under the action of some external force (for example, a pressure gradient, or external electric and magnetic fields).

Here we will focus our attention only on the spontaneous orientational transitions. Due to a very small amount of experimental data on the orientational transitions of the second group, we will not consider them. Only orientational transfers induced by the action of electric and magnetic fields (the Fredericks transition and the flexo-electric effect) will be briefly discussed in Chapter 6, together with the other field-induced phenomena in FSLC films.

5.1.1 EXPERIMENTAL DATA

A number of free-standing films, prepared from some thermotropic nematic substances, may preserve the uniform director alignment over a wide range of film thicknesses (h) and temperatures (T). This is the case of, for example, the nematic 4'-n-penthyl-4-cyanobiphenyl (5CB) films, which have a stable homeotropic texture [1]. Free-standing films made from some other nematic materials exhibit spontaneous orientational transfers with changes of h and T [1–3].

It is convenient to study the orientational transitions in such films using the rectangular frames [3], shown in Fig. 3.2. These frames allow one to obtain the homogeneous director orientation by applying a stabilising electric field, and to change the film thickness (see Chapter 3).

Firstly, we will give several examples of the orientation transitions in nematic films, taking place when h is changing.

(i) At room temperature ($T = 25\,°C$), the nematic MBBA (see Chapter 1) freely suspended film exhibits three types of texture, depending on the h value:

1. the high thickness structure (HTS) observed for $h > 70\,\mu m$, which represents the spatially inhomogeneous tilted orientation of the director, Fig. 5.1(a) [2, 3];
2. the intermediate thickness structure (ITS), which gradually appears when $h < h'_c \approx$ several tens of μm. In the ITS the nematic director is oriented homeotropically (the film looks black in the crossed microscope Nicol prisms), Fig. 5.1(b) [3];
3. the low thickness structure (LTS), which appears abruptly when $h < h''_c \approx 10\,\mu m$. It represents the tilted texture with a great number of disordered black lines with a 'nucleus' in their intersection points (schlieren texture), Fig. 5.1(c) [3].

(ii) At room temperature, a free-standing film of the nematic 4-ethyl-2-fluoro-4'-[2-($trans$-4-n-pentylcyclohexyl)ethyl]-biphenyl (I52) exhibit two orientational states depending on the value of h:

1. in the HTS and ITS ($h = 100$–$10\,\mu m$) the director orientation in such a film is tilted.
2. in the LTS ($h < h_c \approx$ several μm) it gradually changes to a hometropic one [1].

Now let us take several examples of orientational transitions in free-standing nematic films with a change of temperature.

(a) The ITS of an MBBA free-standing film exhibits a gradual orientational transition at $T = T_c \approx 22.2\,°C$ from the initial homeotropic at room temperature, to the tilted orientation when T is lowered. The inverse transition takes place when the film is heated [1, 3].
(b) The ITS of I52 free-standing film exhibits a gradual change of the director orientation at $T = T_c \approx 23.6\,°C$ from the initial tilted at room temperature, to the homeotropic one on cooling. This transition is also reversible [1].

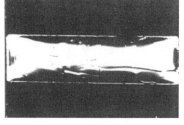

(a)

(b)

(c)

Fig. 5.1 Three types of director orientation observed in free-standing MBBA films (viewed in a polarising microscope) [3] (by permission of Pergamon Press): (a) high thickness structure (HTS); (b) intermediate thickness structure (ITS); (c) low thickness structure (LTS)

For both these orientational transfers, the director tilt angle θ alters homogeneously over the film thickness. Such a change corresponds to a minimal deviation of the film free energy, Fig. 5.2.

Experimental temperature dependencies of θ^2 (the value determined from the polarisation-optical measurements, see Eqs. 3.7 and 3.8) for both MBBA and I52 films are found to be linear for $\theta \ll 1$ [1, 3]:

$$\theta \propto |T_c - T|^{1/2}. \tag{5.1}$$

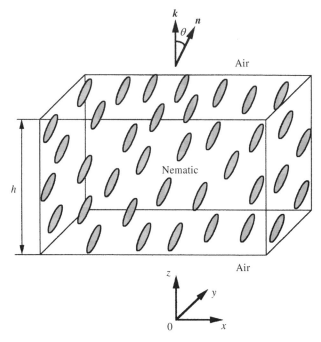

Fig. 5.2 A schematic diagram showing the orientation of the nematic director **n** with respect to the surface normal **k** of a freely suspended film: h is the film thickness (reproduced with permission from Reference [1]. Copyright 1995 by the American Physical Society)

The character of the $\theta(T)$ dependencies allows us to name the considered temperature direction transfers 'second order' orientational transitions.

Note that numerous orientational transfers, similar to that just described, have also been observed in samples of some nematics, bounded from both sides by solid substrates, or only with one free surface. These phenomena occur under changes in the sample thickness, temperature, degree of impurity adsorption at the surfaces, etc. These are often called the 'local' and 'spontaneous' Fredericks transitions (see, for example, [1, 4] and the literature cited therein).

5.1.2 THE GENERALISED PARSONS MODEL

Orientational transitions and the temperature behaviour of the tilt angle of some FSLC films will be discussed in the framework of the following models:

1. the theory of Mada [5, 6], which balances the elastic torques due to distortion of the director field near the free surface with those due to the 'easy orientation' axis at the interfaces.

2. Parsons model [7], which takes into account the competition between quadrupolar and dipolar forces. Quadrupolar forces arise due to van der Waals interactions between the anisotropic molecules of the nematic. They contribute a surface-energy term $(\boldsymbol{n} \cdot \boldsymbol{k})^2$, which favours planar alignment. Polar forces arising from an asymmetry between two ends of the liquid crystal molecule yield an $(\boldsymbol{n} \cdot \boldsymbol{k})$ surface energy term, which generally favours homeotropic alignment.

3. the theory of Barbero *et al.* [8], which introduces an order-electric surface polarisation P_{oe} arising from the spatial variation of the nematic order parameter Q (see Chapter 1) near the boundaries. Their approach is similar to that of Parsons; however, the presence of P_{oe} implies an additional term $\propto (\boldsymbol{n} \cdot \boldsymbol{k})^4$ in the surface free energy. This model does not incorporate a polar term.

4. the generalised Parsons model [1]. This most recent approach is based on the Landau decomposition of the surface tension γ of the free surface of a nematic in the powers of $(\boldsymbol{n} \cdot \boldsymbol{k}) = \cos \theta$ (see Fig. 5.2). It combines the polar and quadrupolar contributions, introduced in the Parsons model, with the order-electric polarisation contribution, introduced in Barbero *et al.* theory:

$$\gamma = \gamma_0 - \gamma_p \cos \theta + \tfrac{1}{2}\gamma_q \cos^2 \theta - \tfrac{1}{4}\gamma_{oe} \cos^4 \theta. \tag{5.2}$$

Here γ_0 is the isotropic component of the surface tension, and γ_p, γ_q and γ_{oe} are the polar, quadrupolar and order-electric components of the surface tension, respectively.

Note that expressions analogous to (5.2) have been used to describe the anchoring of nematics with respect to solid surfaces (e.g. [4]).

The Landau coefficients γ_0, γ_p, γ_q and γ_{oe} in Eq. (5.2) are allowed to vary with different thermodynamic control parameters (temperature, ion concentration, etc.). Let us identify the physical mechanisms that determine these coefficients.

The polar and quadrupolar coefficients of the surface tension γ_p and γ_q contain contributions from a number of physical effects. One can write

$$\gamma_p = \gamma_{st} \pm \gamma_{dp} \tag{5.3}$$

and

$$\gamma_q = \gamma_{vdW} \pm \gamma_{sp} \pm \gamma_{fe} - \gamma_{sm} + \gamma_{oe}. \tag{5.4}$$

Here, the indices 'st', 'dp', 'vdW', 'sp', 'fe', 'sm' and 'oe' refer, respectively, to effective surface terms coming from steric forces, orientation of the surface dipole moments, van der Waals (dispersion) forces, surface polarisation, gradient flexoelectricity, smectic ordering, and order-electric contributions. The '+' and '−' signs for γ_{dp}, γ_{sp} and γ_{fe} correspond to longitudinal and transverse molecular dipole moments, negative and positive dielectric anisotropies ε_a and negative and positive flexoelectric coefficients e, respectively.

Let us discuss in more detail the physical nature of each of these contributions.

γ_{st}: This contribution may appear when the nematic molecules have the anisotropy of form (e.g. different end groups). If one of these ends prefers to stay in the bulk of the mesophase, homeotropic alignment will be stabilised [7].

γ_{dp}: This contribution may appear for nematics whose molecules have constant (longitudinal or transverse) dipole moments. These molecular dipoles are usually aligned in one direction near the surface. They form the surface layer with a macroscopic polarisation P_{dp} and thickness $l_s \approx 10^{-6}$ cm, where l_s is the diffusion length [7, 9].

γ_{vdW}: The van der Waals contribution is due to the fluctuation-induced dipole–dipole interactions. It is present even in the absence of permanent dipole moments. The interaction of an individual molecule with the surrounding medium near a surface falls with the distance z as z^{-3}. Van der Waals forces usually orient the liquid crystalline director parallel to the free surface [10].

γ_{sp}: This contribution is due to the impurity ions which are adsorbed at the nematic film surfaces and form double ion layers. The double layer results in a surface ionic polarisation P_i and hence in an electric field perpendicular to the boundaries [11, 12]. The strength E_s of this field decreases exponentially as the distance from the surface increases, with a characteristic decay radius L_D, the Debye screening length [13]. The contribution γ_{sp} to the surface tension represents the aligning effect of the surface electric field E_s upon the liquid crystalline director. Depending on the sign of the mesophase dielectric anisotropy, this surface term can stabilise either homeotropic or planar orientations [11, 12]. Note that the electric field created by the surface molecular dipoles, which results in γ_{dp}, can also contribute to the field E_s induced by the adsorbed ions, and hence to γ_{sp}. Experiments on the electrokinetic effect in the isotropic phase of some thermotropic nematics find for the surface-adsorbed ion a charge density $\sigma \approx 10^{-12}$ C/cm^2 [14]. The surface polarisation P_s was also detected by means of optical second-harmonic generation [15] and modulation spectroscopy of light reflection at the interface between nematics and semiconductors. These last measurements gave $E_s \approx 10^{-4}$ V/cm and $L_D \approx 10^{-5}$ cm [16, 17]. However, neither of these methods could distinguish between the contributions to P_s coming from P_i and from P_{dp}.

γ_{fe}: Since the surface electric field E_s created by surface adsorbed ions or molecular dipoles is spatially inhomogeneous, the gradient flexoelectric effect may also be important. This effect results in the contribution γ_{fe}. Depending on the sign of the flexoelectric coefficient e of the nematic substance, this contribution may stabilise either homeotropic or planar orientation [18].

γ_{sm}: The smectic ordering near the surfaces of nematic liquid crystals already described in detail in the Chapter 4 leads to the surface energy term γ_{sm}, which stabilises the homeotropic director orientation.

γ_{oe}: As has just been mentioned, the order-electric polarisation P_{oe} appears due to spatial gradients of the nematic order parameter ∇Q (see also Chapter 1), which exist over distances from the surface that are comparable with the nematic correlation length $\xi \approx 5 \times 10^{-6}$ cm. This results in the contribution γ_{oe}, [8, 19].

This term can be important in cases where ∇Q is large, e.g. at temperatures near the nematic–isotropic transition point T_{NI} [20]. Since the experiments considered in the previous paragraph were done at temperatures far from T_{NI}, ∇Q should be small, so that this is a small contribution ($\gamma_{oe} < \gamma_p$, γ_q). As a result, only terms proportional to $\cos\theta$ and $\cos^2\theta$ need be considered in Eq. (5.2).

To find the equilibrium director orientation in the nematic free-standing film, one should minimise Eq. (5.2) with respect to the tilt angle θ (the $\cos^4\theta$ term is omitted):

$$\sin\theta(\gamma_p - \gamma_q\cos\theta) = 0. \qquad (5.5)$$

Eq. (5.5) has two equilibrium solutions

$$\theta = 0 \qquad \text{(stable for } \gamma_p > \gamma_q), \qquad (5.6)$$

$$\cos\theta = \frac{\gamma_p}{\gamma_q} \qquad \text{(stable for } \gamma_p < \gamma_q). \qquad (5.7)$$

For small angles of θ (near T_0) where $\cos\theta$ can be expanded as $\cos\theta \approx 1 - \theta^2/2$, the temperature dependence of θ for ($\gamma_p < \gamma_q$) can be expressed as

$$\theta = (A/B)^{1/2}, \qquad (5.8)$$

where $A = A_0|T - T_0| = \gamma_p - \gamma_q$ (A_0 is a constant) and $B = 1/2\gamma_q$ are the Landau coefficients.

It is evident that Eq. (5.8) has the form of the experimentally observed $\theta(T)$ dependence (see Eq. 5.1).

Now let us try to apply the just-elaborated generalised Parsons model to some concrete nematic materials.

5CB: For this substance $\varepsilon_a > 0$ and $e > 0$, $\gamma_p = \gamma_{st} + \gamma_{dp}$ and $\gamma_q = \gamma_{vdW} - \gamma_{sp} - \gamma_{fe} - \gamma_{sm}$. Because of the longitudinal molecular dipole moment and the considerable smectic surface ordering, one should expect $\gamma_p \gg \gamma_q$. This means that the homeotropic director orientation will be stable, which agrees with the experimental observations (see the previous paragraph).

MBBA: For this material $\varepsilon_a < 0$ and $e < 0$, $\gamma_p = \gamma_{st} - \gamma_{dp}$ and $\gamma_q = \gamma_{vdW} + \gamma_{sp} + \gamma_{fe}$ (transverse molecular dipole moment, no smectic surface ordering). One should expect that the coefficients γ_p and γ_q in this case have the same order of magnitude, and that γ_q is strongly temperature dependent, due to the temperature dependence of the coefficients ε_a and e, which are contained in the terms γ_{sp} and γ_{fe}, respectively. Then, it is possible that for $T < T_0$ $\gamma_p < \gamma_q$, i.e. tilted orientation is stable; at $T = T_0$ $\gamma_p = \gamma_q$, i.e. an orientational transition takes place; and for $T > T_0$ $\gamma_p > \gamma_q$, i.e. the homeotropic orientation is stable. This behaviour corresponds to the experimental observations (see the previous paragraph).

I52: For this nematic $\varepsilon_a < 0$ and $e < 0$, $\gamma_p = \gamma_{st} - \gamma_{dp}$ and $\gamma_q = \gamma_{vdW} + \gamma_{sp} + \gamma_{fe} - \gamma_{sm}$ (transverse molecular dipole moment, surface smectic ordering). At higher temperatures, well above T_{NA}, the smectic-ordering term γ_{sm} is negligible. So, one expects $\gamma_p < \gamma_q$ and the tilted director alignment. As the temperature is decreased and the smectic-ordering term becomes more important, an orientational transition takes place at $T = T_0$, where $\gamma_p = \gamma_q$. For $T < T_0$, $\gamma_p > \gamma_q$ and the orientation is homeotropic. This behaviour agrees with the experimental data (see the previous paragraph).

Since some of the coefficients, γ, in Eq. (5.2) depend on the film thickness, h (for instance, γ_{sp}, γ_{oe} [1]), the total surface energy will also be h-dependent. This explains qualitatively the just-described orientational transitions induced by the change in h. Note that such orientational transfers are similar in their physical mechanism (a balance between contributions into the free energy from different surface forces) to the spontaneous [21, 22] and local [23–25] Fredericks transitions already mentioned in the previous paragraph. These are in fact orientational transitions in nematic layers (in contact with a solid substrate) due to the variation of control parameters such as the layer thickness, temperature, etc.

5.2 AN EXAMPLE OF ORIENTATIONAL TRANSITION IN SMECTIC C* FILMS

The freely suspended smectic C* films prepared from substances exhibiting the temperature-induced inversion of the sign of the spontaneous polarisation, P_{sp} (see, for example, [9]), were studied by means of ellipsometry in Reference [26]. The initial orientation of P_{sp} was fixed by the externally applied electric field. The inversion of the P_{sp} sign was observed by the reversal of the director tilt. It took place in free-standing films at a given temperature, T_i, over the whole film thickness, h. T_i was found to decrease linearly with diminishing h, indicating the influence of the reduced free-standing film dimensionality.

5.3 PHASE TRANSITIONS

Phase transitions in FSLC films have a lot of specific features, compared to those in thick, bulk mesophase samples.

As we already know, a free-standing film may be considered either a three-dimensional, or a two-dimensional, depending on concrete experimental conditions (see Chapter 2). For three-dimensional films, the influence of the film surfaces, which usually stabilise mesophases with lower symmetry than those constituting the film volume (see, for example, [4]), makes phase transitions more variable than in bulk mesophases. In two-dimensional (*xy-*) films, phase transfers also exhibit a number of interesting peculiarities: the appearance of new phases, the important role of thermal fluctuations (which should, in principle, stabilise a higher-symmetry phase in thin films), etc.

Consider now in more detail the phase transition phenomena in three- and two-dimensional FSLC films: firstly, basic theoretical predictions obtained in the framework of the mean field (Landau–de Gennes) theory, and secondly, some of the experimental data.

5.3.1 THEORY: THREE-DIMENSIONAL FILMS—ROLE OF INTERFACES

To illustrate the symmetry-lowering role of a surface, let us analyse the two following easy examples of phase transitions in semi-infinite (three-dimensional) liquid crystalline samples:

1. nematic—isotropic phase transition [4, 27, 28];
2. smectic C* (C)—smectic A phase transition [4, 29].

Note that the further result obtained can easily be modified for the case of two boundaries (or interfaces).

1. The free energy per unit area of a nematic sample can be written as [28]

$$\Phi_1 = \int_0^\infty \left[F_0(Q) + K\left(\frac{dQ}{dz}\right)^2 \right] dz - GQ_s. \qquad (5.9)$$

Here $F_0(Q) = a(T - T^*)Q^2 + bQ^3 + cQ^4$ is the uniform part of the free energy, a, b, c, and T^* being the coefficients in Landau's expansion; K is the modulus of elasticity (in the one-constant elastic approximation)—see Chapter 1; G is a positive constant characterising the degree of anchoring at the surface; Q is the nematic scalar order parameter (see Chapter 1), $Q_s = Q(z = 0)$ is the surface value of Q ($z = 0$ is the co-ordinate of the surface).

It is obvious that the bulk part of Eq. (5.9) stabilises the higher-symmetry isotropic phase ($Q = 0$, ∞ symmetry), while the surface part stabilises the lower-symmetry nematic phase ($Q \neq 0$, ∞/mm symmetry).

Consecutive minimisation of Eq. (5.9) with respect to $Q(z)$ and Q_s, taking into account the additional boundary condition

$$\left.\frac{dQ}{dz}\right|_{z \to \infty} = 0, \qquad (5.10)$$

leads to the following equation:

$$\xi^2 \left(\frac{dQ}{dz}\right)^2 = \varphi(Q) - \varphi(Q_b). \qquad (5.11)$$

Here $\xi = (K/aT^*)^{1/2}$ is the correlation length, $\varphi(Q) = F_0(Q)/aT^*$ and Q_b is the bulk value of the order parameter.

Substituting this integral into Eq. (5.9), we obtain for the energy $\mathscr{F}_{sl}$ of the surface layer

$$\frac{\mathscr{F}_{sl}}{\xi aT^*} = 2 \int_{Q_b}^{Q_s} [\varphi(Q) - \varphi(Q_b)]^{1/2} dQ - gQ_s, \qquad (5.12)$$

where $g = G/\xi aT^*$ is the dimensionless surface potential.

This equation has many solutions, but the concrete value of Q_s is that which minimises $\mathscr{F}_{sl}$. Once Q_s has been determined, the distribution of the order parameter $Q(z)$ can be calculated from Eq. (5.11):

$$\frac{z}{\xi} = \int_{Q(z)}^{Q_s} \frac{dQ}{[\varphi(Q) - \varphi(Q_b)]^{1/2}}. \qquad (5.13)$$

This equation can be solved numerically. Some of the results of the calculations of $Q(z)$ are shown in Fig. 5.3. It is clearly seen from this Figure that the surface induces nematic ordering even when all of the bulk is already in the isotropic state.

2. The free energy per unit area of a smectic C* sample can be written in the one-constant elastic approximation as [29]

$$\Phi_2 = \int_0^\infty \left[\frac{K}{2} \left(\frac{d\Theta}{dz} \right)^2 + f(\Theta) \right] dz + f_0(\Theta_0). \qquad (5.14)$$

Here K is the mean elastic constant, θ is the smectic C* two-component order parameter (see Eq. 1.7), $\Theta_0 = \Theta(z = 0)$ is the surface value of the order parameter

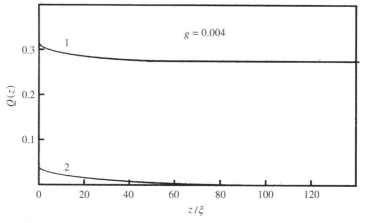

Fig. 5.3 Theoretical dependencies of the order parameter ($Q(z)$) upon the distance from the surface, calculated for nematic 5CB: (1) $T \le T_{NI}$; (2) $T \ge T_{NI}$ ($|T - T_{NI}| \approx 0.1\,^\circ C$), T_{NI} being the temperature of the bulk nematic–isotropic phase transition (reprinted with permission from Reference [28]. Copyright 1982 by the American Physical Society)

($z=0$ is the co-ordinate of the surface) and the function $f(\theta)$ and $f_0(\theta_0)$ can be written as follows:

$$f(\Theta) = -\lambda\left(\Theta_1 \frac{d\Theta_2}{dz} - \Theta_2 \frac{d\Theta_1}{dz}\right) - \frac{A}{2}(\Theta_1^2 + \Theta_2^2)$$

$$+ \frac{B}{4}(\Theta_1^2 + \Theta_2^2)^2, \qquad (5.15)$$

$$f_0(\Theta_0) = \frac{1}{2}\sigma\Theta_0^2 - \boldsymbol{H}\Theta_0. \qquad (5.16)$$

Here Θ_1 and Θ_2 are the components of the order parameter Θ (see Eq. 1.7); λ is the coefficient of the Lifshits invariant, which characterises the rotation of the director when one moves from one smectic layer to another; A and B are the Landau expansion coefficients ($A \approx T_{CA} - T$, where T_{CA} is the temperature of the smectic C*–smectic A bulk phase transition); coefficient σ characterises the anchoring energy; and the vector-coefficient $\boldsymbol{H}(H_1, H_2)$ is due to the dipole–dipole surface interaction (the smectic C* molecular dipoles are oriented in one preferred direction at the surface).

It is obvious that the surface free-energy term (Eq. 5.16) stabilises the lower symmetry smectic C* phase: $\Theta \neq 0$, $\infty 2$ point symmetry group.

Minimisation of Eq. (5.14) with respect to Θ leads to the following equilibrium equation and boundary condition:

$$K\frac{d^2\Theta}{dz^2} = \frac{\partial f(\Theta)}{\partial \Theta}, \qquad (5.17)$$

$$K\frac{d\Theta}{dz} = \frac{\partial f_0(\Theta_0)}{\partial \Theta_0}, \qquad z = 0. \qquad (5.18)$$

From Eq. (5.17), for small polar angles θ ($\theta \ll 1$) and undisturbed by the surface smectic C* helix ($H_2 = 0$ and $\phi = q_0 z$, where $q_0^{-1} = K/\lambda$ is the helix pitch), one has:

$$\frac{K}{2}\left(\frac{d\theta}{dz}\right)^2 - f(\theta) = -f(M). \qquad (5.19)$$

Here the following relation was taken into account:

$$\lim_{z \to \infty} \theta(z) = M, \qquad (5.20)$$

Eq. (5.19) gives the following profile of the order parameter:

$$\sqrt{K}\frac{d\theta}{dz} = \pm[2f(\theta) - 2f(M)]^{1/2}, \qquad (5.21)$$

and, for the surface angle θ_0, we obtain

$$\frac{1}{\sqrt{K}}\frac{\partial f_0(\theta_0)}{\partial \theta_0} = \pm[2f(\theta_0) - 2f(M)]^{1/2}. \qquad (5.22)$$

Here ' + 'corresponds to $\theta_0 < M$ and ' − ' to $\theta_0 > M$.

Eq. (5.22) gives several solutions for θ_0. To choose a single one, we should take such a θ_0 which gives the minimum to the total free energy of our system (see Eq. 5.14):

$$\sigma\theta_0 - H_1 = \frac{1}{M}\left(\frac{A'K}{2}\right)^{1/2}(M^2 - \theta_0^2). \qquad (5.23)$$

Here $A' = A + \lambda^2/K$.

Note that, in addition, $\theta_0 < M$, since $H_1 \ll \sigma$ (experimentally $H_1 = 10^{-2}\,\sigma$ [4, 29, 30]).

Taking into account Eq. (5.23) and the relation $A' = BM^2$, obtained from the minimisation of the bulk energy, we get the following equation for $\theta(z)$:

$$\frac{d\theta}{dz} = \frac{1}{M}\left(\frac{A'}{2K}\right)^{1/2}(M^2 - \theta^2). \qquad (5.24)$$

Now we can apply to Eq. (5.24) and the boundary condition (which follows from Eq. 5.18)

$$K\frac{d\theta}{dz} = \sigma\theta_0 - H_1, \qquad z = 0 \qquad (5.25)$$

the graphical approach, similar to that elaborated in References [31–33]. The graphical solution of Eqs. (5.24) and (5.26) is represented in Fig. 5.4. It is obvious that for $\sigma < \sigma_c = (2A'K)^{1/2}$ there are two points of intersection of the straight line (5.25) with the parabola (5.24), which corresponds to the first order phase transition. For $\sigma > \sigma_c$ there is only one intersection point, i.e. the second order phase transition. To evaluate σ_c, we will take $M \approx 10^{-2}\,$rad, (near T_{CA}), $B \approx 10^7\,$erg/cm^3 (B is the modulus of compressibility of smectic layers), $K \approx 10^{-6}\,$dyne then, using the relation $A' = BM^2$, we will obtain $\sigma_c \approx 10^{-2}\,$erg/cm^2, a result which corresponds to the experimental values of the anchoring energy (see Chapter 2).

The problem of the first-order phase transition can be solved numerically only, using graphical integration for evaluation of the free energies (see, for example, [34]). Here we will consider only the case of quite strong anchoring, where a second order phase transfer is realised.

Integration of Eq. (5.24) gives the following profile for the order parameter $\theta(z)$:

$$\theta(z) = M\tanh\left(\frac{A'}{2K}\right)^{1/2}(z + z_0), \qquad (5.26)$$

where z_0 is a constant determined by the formula (5.23).

Eq. (5.26) gives two equilibrium values of θ_0 (negative and positive), but only the positive minimises the free energy (5.9). In any case, θ_0 should be positive as the modulus of the order parameter. Then $\theta(z) > 0$ and $z_0 > 0$.

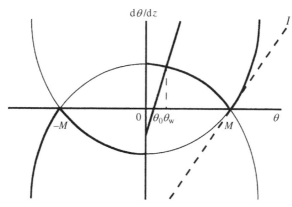

Fig. 5.4 Graphical solution of the equilibrium equation (5.24) together with the boundary condition (5.25). The line I divides the areas with the first and second order phase transitions (adapted from Reference [29])

It is obvious that with growing temperature ($A' \to 0$) the point M in Fig. 5.4 will move in the direction of the point $\theta_w = H_1/\sigma$ which corresponds to $d\theta/dz|_{z=0} = 0$. At some temperature $T = T_w$, $M = \theta_w = \theta(z) = \theta_0$, z_0 diverges logarithmically and the order parameter in the bulk of the mesophase will be determined only by the surface. This means that a macroscopic layer of the smectic C* paraphase (smectic C* phase with quite low $\theta \approx 10^{-2}$ rad) is formed near the surface, i.e. a second order wetting transition takes place.

From the relation $A' = BM^2$ we have for T_w

$$T_c - T_w \approx B(H_1/\sigma)^2. \tag{5.27}$$

The profiles of $\theta(z)$ for $T < T_w$ and $T = T_w$ are schematically represented in Fig. 5.5.

Note that the first order phase transition smectic C*–smectic C* paraphase, which we have not considered here in detail, should also be of the wetting type (see, for example, [4, 29]).

In the case of the phase transition smectic C–smectic A, the second term in Eq. (5.16) will be zero, since there is no spontaneous polarisation. This means that $H_1 = 0$, $A' = A$ and $\theta(z) = 0$ for $T = T_c$, i.e. the smectic C phase will not now be stable at the surface, and the phase transitions in the bulk and near the surface will take place at the same temperature (there is no wetting behaviour).

5.3.2 THEORY: TWO-DIMENSIONAL FILMS

Consider now the main characteristics of phase transitions in thin (two-dimensional) FSLC films. Firstly, some comments concerning the thermodynamic stability of such films. The presence of thermodynamic fluctuations of crystalline order is a universal property: in solid crystals the positional fluctuations prevail, while in liquid crystals

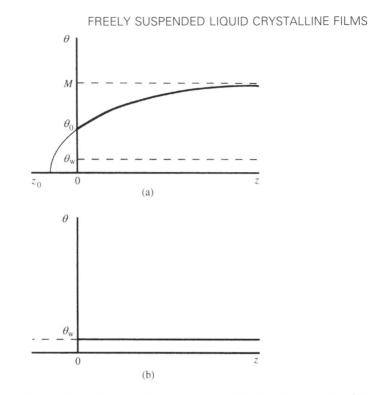

Fig. 5.5 Schematic profiles of the order parameter $\theta(z)$ for the smectic C*– smectic C* paraphase wetting transition: (a) $T < T_w$; (b) $T = T_w$ (adapted from Reference [29])

the orientational fluctuations are dominant (see, for example, [35, 36]). It has been shown (e.g. [37, 38]) that thermodynamic fluctuations in a crystalline or liquid crystalline xy-system diverge logarithmically as the size r of a system increases. This is the so-called Landau–Peierls divergence. This is why only two-dimensional films with restricted sizes $r < \zeta \propto \exp(T^{-1})$ (where ζ is the characteristic correlation size of the system at which the system will be destroyed by the fluctuations) can be thermodynamically stable. Note that thermodynamically stable two- or one-dimensional systems of unlimited size may exist only at $T = 0\,\mathrm{K}$, for which $\zeta = \infty$.

Phase transitions in two-dimensional FSLC films are described by the changes in corresponding order parameters. For analysis of the nematic–isotropic phase transfer one should now use the order parameter Q_{xy} of the form of Eq. (1.2), written for one molecular layer. Then, $Q_{xy} \neq 0$ in the nematic phase, and $Q_{xy} = 0$ in the isotropic phase.

Note that for two-dimensional smectics, the order parameter $|\Psi|$ in the form of (1.6) has no physical meaning, since, for such films, there are no density modulations in the direction perpendicular to the xy-plane.

For two-dimensional free-standing smectic C* (or smectic C) films, it is convenient to take for the order parameter the projection c of the director in the plane of the smectic layers, i.e. the tilt angle θ (for $\theta \ll 1$), see Eq. (1.7) and Fig. 1.2e. If one considers the smectic C* (smectic C)–smectic A phase transition in such a film, then $\theta \neq 0$ for the smectic C*(C) phase and $\theta = 0$ for the smectic A phase (see, for example, [39–41]).

Finally, analysing the phase transfers involving the smectic phases with structured layers, one should take into account the BO ordering (see Eqs. 1.8 and 3.15) and Fig. 1.7b).

It has been shown that the $6n$-fold BO order parameters should obey the following relation in the critical region [42]:

$$C_{6n} = C_6^{\sigma_n}, \tag{5.28}$$

where $\sigma_n = n + x_n n(n-1)$ with $x_n \approx 0.3$ for a three-dimensional system, and $x_n = 1$ for a two-dimensional one.

Let the molecular layers of a thin tilted smectic free-standing film (and the film itself) lie in the (xy) co-ordinate plane. Then it is convenient to understand under the two-dimensional liquid crystalline director the unit vector (see Fig. 1.7)

$$c = i \cos \varphi + j \sin \varphi, \tag{5.29}$$

where i and j are the unit vectors lying, respectively, along the x and y co-ordinate axes; φ is the azimuthal angle co-ordinate of the bulk liquid crystalline director n.

Vector c coincides in its direction with the projection of n into the plane (xy) of the film. Thus the director distribution in the film plane will be determined only by the azimuthal angle $\phi(x, y)$, and c will serve as the vector order parameter.

Two-dimensional liquid crystalline films are interesting objects with which to test the applicability of the general Kosterlitz and Thouless (e.g. [43–45]) model of the phase transition in two-dimensional space. It predicts that the xy-system will melt via a defect binding–unbinding phase transition. This phenomenon has several peculiar features when compared to a standard second-order phase transfer. The thermodynamic variables of the system scale as powers of the correlation length ξ_+ (defined as the length over which orientational order persists), but the correlation length itself does not scale as a power of the reduced temperature. Instead its temperature dependence has an essential singularity at the phase transition temperature T_{xy}:

$$\xi_+ = \xi_0 e^{b/|T - T_{xy}|^{1/2}} \tag{5.30}$$

for $T \rightarrow T_{xy}$ from above T_{xy} in temperature, where ξ_0 has the order of the film thickness; for $T < T_{xy}$, $\xi_+ = \infty$.

The elasticity of an xy-liquid crystalline film is characterised by two two-dimensional elastic constants $K_\perp$ and $K_\parallel$ measured in the directions in the film plane perpendicular and parallel, respectively, to the in-plane projection of the director. If, for some xy-mesomorphic film, one approaches the temperature T_{xy} of the phase transition into the higher-symmetry phase from below, the two-dimensional elastic constants should reach a universal constant value:

$$\lim_{T \to T_{xy}^-} K^*(T) = 2k_b T_{xy}/\pi, \qquad (5.31)$$

where $K^* = \frac{1}{2}(K_\perp + K_\parallel)$, and turn abruptly to zero at $T > T_{xy}$ [41, 45, 46].

The free energy density of any two-dimensional liquid crystalline film will be determined by the constants $K_\perp$ and $K_\parallel$. For example, for a two-dimensional smectic C* film it has the following form [41, 46]:

$$F = \frac{1}{2}K_\perp \left(\frac{\partial \varphi}{\partial r_\perp}\right)^2 + \frac{1}{2}K_\parallel \left(\frac{\partial \varphi}{\partial r_\parallel}\right)^2 - P_o h E_a \cos \varphi + F_{d-d}. \qquad (5.32)$$

Here r is the radius-vector in the film plane; $\varphi(r)$ is the azimuthal co-ordinate of the liquid crystalline director; P_a is the amplitude of the in-plane polarisation of the ferroelectric liquid crystal; h is the film thickness; E_a is the amplitude of an aligning electric field; and F_{d-d} is the dipole–dipole term.

The two-dimensional elastic constants, which correspond to the restoring force in response to deformations $\varphi(r)$ perpendicular to the direction of the average tilt, can be given in terms of the bulk splay, twist and bend elastic constants K_{11}, K_{22} and K_{33} as [41, 46]

$$K_\perp = hK_{11} \sin^2 \theta, \qquad (5.33')$$

$$K_\parallel = h \sin^2 \theta(K_{22} \cos^2 \theta + K_{33} \sin^2 \theta) = \bar{K}h \sin^2 \theta. \qquad (5.33'')$$

Then, using Eq. (5.31), one can calculate the universal jump value in K^*

$$2k_b T_{xy}/\pi = \frac{1}{2}(K_{11} + \bar{K})h \sin^2 \theta^*, \qquad (5.34)$$

where θ^* is the predicted jump in θ.

Note, finally, that the phase diagrams of two-dimensional films should, in principle, differ from those of three-dimensional samples. Some theoretical models (see, for example, [44, 47, 48]) prove, for instance, the possibility of the appearance in xy-smectic films, just above the melting point, of hexatic mesophases. Different periodical (the so-called modulated) two-dimensional equilibrium phases are also predicted theoretically for such films (e.g. [49–53]).

More information on phase transitions in two dimensions can be found, for example, in the review article Reference [54].

5.3.3 EXPERIMENTAL DATA

A great number of publications concerning the experimental studies of different phases and phase transitions in FSLC (especially, thermotropic smectic) films has appeared during the last several years. As a consequence, it will be quite difficult to make a comprehensive review of all the actual trends in this field, so, we will give here only some of the most interesting (in our opinion) examples of experimental work, especially those illustrating the above-considered theoretical statements. The interested reader can find more material and references in, for instance, References [55–58].

The experimental evidence of the just-described lowering-symmetry effect of the interfaces in freely suspended nematic or isotropic films (disordered micellar phase, for example), is the stratification phenomenon observed for $h < 1000$ Å. It proves that a surface may impose the lower-symmetry nematic (∞/mm) or smectic-like ($\infty 2$) ordering in isotropic and nematic films (see Chapter 4).

The experimental optical studies of the above theoretically analysed phase transition between the oriented (by an electric field) smectic C* and smectic A phases in free-standing films (e.g. [39–41, 46, 59–61]), show that the character of phase transfer and its temperature (T_{CA}) depend strongly on the film thickness (the number N of smectic layers in the film). For example, the ellipsometric data for 4-(3-methyl-2-chloropentanoyloxy)-4'-heptyloxybiphenyl) (C7) free-standing films indicate that the smectic C*–smectic A phase transition is of first order in thick (bulk, three-dimensional) films, while it becomes continuous for films with $N < 15$ (two-dimensional films) [59]. If the thickness of the C7 free-standing film diminishes even further—until it reaches one bilayer—there will be no transfer from the smectic C* to the smectic A phase on heating. A considerable director tilt angle ($\langle\theta\rangle \approx 30°$) is present for such a film. This is evidence of the extremely strong symmetry-lowering effect of the film interfaces [59].

The ellipsometric studies of freely suspended films prepared from another material, a chiral thermotropic substance p-(n-decyloxybenzylidene)-p-amino-(2-methylbutyl)cinnamate (DOBAMBC), show a smooth variation of the average over the film thickness director tilt angle $\langle\theta\rangle$ with temperature for all N [60] (Fig. 5.6). The elastic constants exhibit similar behaviour in this case [60]. However, the ellipsometric and the light scattering data [46] show that, for the DOBAMBC free-standing films containing only several molecular monolayers, the smectic C*–smectic A phase transition, in the very vicinity of T_{CA} (closer than 1 °C), proceeds with a universal jump in the tilt angle of the two surface monolayers $-\theta_s$, described quite well by Eq. (5.34). The temperature behaviour of the scattering intensities and decay times for DOBAMBC films also indicates the realisation of the xy-surface phase transfer. The light scattering data for temperatures above T_{CA} are in agreement with theoretical predictions for two-dimensional smectic C*(C)–smectic A phase transitions [62]. For both types of films just mentioned, T_{CA} depends strongly on N, and the film surfaces stabilise the lower-symmetry (smectic C*) phase.

The examples just described of the influence of the change of the FSLC dimensionality (i.e. the transfer from thick, three-dimensional, to thin, two-dimensional, films) on the character of the smectic C*–smectic A phase transitions are not unique. For instance, X-ray measurements show that in thick (bulk) free-standing films of n-butyl-4'-n-hexyloxybiphenyl-4-carboxylate (46OBC) the smectic B–

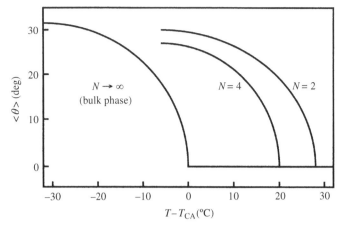

Fig. 5.6 Schematic average director tilt angle dependencies upon temperature as the function of the number of smectic layers (N) contained in the film: T_{CA} is the temperature of the smectic C*–smectic A phase transition for the bulk sample (reproduced with permission from Reference [60]. Copyright 1984 the American Physical Society)

smectic A phase transition is weak first-order, while in a bilayer film of this substance the same phase transfer becomes second-order [63].

The obtained experimental data allow one to propose a schematic model of the molecular organisation inside the smectic C*(C) film in the vicinity of the transition to the smectic A phase, Fig. 5.7. According to this model, only two surface monolayers are in the smectic C*(C) state, i.e. they have the non-zero value of the order parameter θ. All of the resting bulk portion of the film is already in the smectic A phase with $\theta = 0$. It is evident that the smectic C*(C)–smectic A phase transition for such a film may be considered as two-dimensional, since it concerns only the surface monolayers.

There exist also some experimental indications [46, 64] that the inverse phase transfer (from the smectic A to the smectic C* phase) may possibly be realised through the wetting mechanism, similar to that analysed above in this chapter: the surface smectic C*(C) layers gradually increase their thickness h_s with lowering temperature, and at some value $T(= T_w \approx T_{CA})$, h_s may diverge.

The transition between the smectic I and the smectic C phases observed in 4-n-heptyloxybenzylidene-4-n-heptylaniline (7O.7) freely suspended films is another illustration of the lowering-symmetry role of the surface. Above $\sim 78\,^\circ$C, the entire film is in the smectic C state and below $\sim 69\,^\circ$C, the entire film has smectic I structure. Between $69\,^\circ$C and $78\,^\circ$C, the surface layers of the film are in the smectic I state, while the interior layers exhibit the smectic C phase [65]. A similar character of the smectic I–smectic C phase transition was found for free-standing films of the same material by X-ray diffraction [66, 67] and for DOBAMBC films by the calorimetric measurements [68, 69].

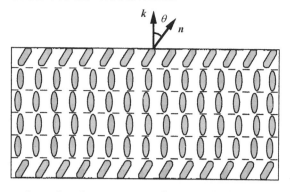

Fig. 5.7 Schematic molecular structure of a thin FSLC film in the vicinity of the smectic C*(C)–smectic A phase transition: The lower-symmetry smectic C*(C) phase is stable near the film interfaces only in two surface monolayers, while the higher-symmetry smectic A phase is stable in the bulk of the film (reproduced by permission from Reference [60]. Copyright 1984 by the American Physical society)

The lowering-symmetry role of the surface has also been found by calorimetric and electron diffraction studies for the smectic A–crystalline B (three-dimensional hexatic phase) phase transition in N-4-n-butyloxybenzylidene-4-n-octylaniline (4O.8) freely suspended films. In this case, the film interfaces stabilise the crystalline B phase [70–72].

A lowering-symmetry effect of surfaces is also manifested during the smectic A–isotropic phase transition in free-standing films. Indeed, the presence of the smectic A bilayers is observed by small-angle X-ray scattering at free surfaces of 4′-n-dodecyl-4-cyanobiphenyl (12CB) samples, heated deep into the isotropic phase $(T - T_{AI} \approx 10\,°\text{C}$, where T_{AI} is the bulk smectic A–isotropic phase transition temperature) [73].

A quite interesting and unusual melting smectic A–isotropic phase transfer (so-called thinning-layer transition) occurs in free-standing films prepared from some fluorinated mesomorphic compounds [74]. As the film temperature is increased to the bulk smectic A–isotropic phase transition point, T_{AI}, the film does not break but starts to thin in a stepwise manner, i.e. the stratification phenomenon takes place (see also Chapter 4). Calorimetric and optical reflection studies show that the exterior boundary areas of the film preserve the smectic A ordering for $T > T_{AI}$. With growing temperature, the consecutive smectic A layers melt, starting from the interior of the film, and squeeze out into the surrounding fluid meniscus, producing in this way a stepwise thinning of the film. The experimental temperature dependence of the film thickness was found to follow a simple scaling law [74]

$$h(t) = l_0 t^{-\nu}, \tag{5.35}$$

where $t = [T_c(N) - T_0]/T_0$, $T_c(N)$ is the maximal temperature at which the N layer film exists, l_0, ν and T_0 are the fit constants having the following values: $l_0 = 0.34 \pm 0.02$, $\nu = 0.74 \pm 0.02$, $T_0 = 84.84 \pm 0.04\,°\text{C}$.

In Ref. [75], the just-described layer-thinning transition is explained by the appearance (for $T > T_{AI}$) of a quasi-smectic A phase inside the film, at the border between the interfacial smectic I and the interior smectic A phases. In the quasi-smectic A phase, the orientational order parameters are different from zero but decay very rapidly to nearly zero with distance. The quasi-smectic A layers easily squeeze from the film into the surrounding meniscus, producing film stratification.

In Reference [75], in contrast, this layer-thinning transfer is associated with the existence of a number of metastable energetic states for a free-standing smectic film, between which it passes with changing temperature.

Phase diagrams of thin freely suspended films (with a thickness of less than 1000 Å) are generally richer in detail than those of the bulk samples of the same substances. In particular, some new hexatic phases predicted theoretically (see above in this chapter) may appear. For example, X-ray studies of the above-mentioned mesomorphic material 7O.7 show that in the temperature range between 30 and 85 °C there are 11 distinct bulk phases. In thin free-standing films of this compound four additional phases appear: two crystalline B phases (hexagonal ABC and orthorhombic-*a*) and two liquid crystalline phases (smectics F and I, see Fig. 1.3) [65, 77–79], Fig. 5.8.

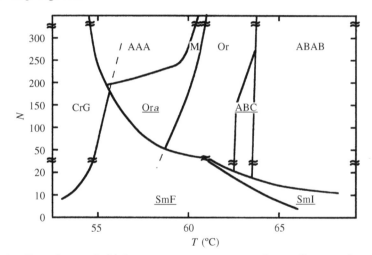

Fig. 5.8 Experimental thickness vs. temperature phase diagram for the 7O.7 free-standing film. The solid lines indicate the reversible phase boundaries. The dashed (dotted) lines indicate the phase boundaries of the metastable phases observed kinetically on heating (cooling). The phases are crystalline G (CrG), smectic F (SmF), smectic I (SmI) and the crystalline B (CrB) modifications: hexagonal closed-packed ABAB (ABAB), hexagonal ABC (ABC), orthorhombic (Or), orthorhombic-*a* (Or*a*), monoclinic (M) and hexagonal AAA (AAA). The smectic C phase (SmC) occurs above about 69 °C. The names of new phases ABC, Or*a*, SmF and SmI, appearing only in thin films, are stressed (reproduced with permission from Reference [79]. Copyright 1987 by the American Physical Society)

Schematic structures of different crystalline B phases of 7O.0 are represented in Fig. 5.9. The crystalline B phases are three-dimensional crystals with the director oriented normal to the layers. The in-plane nearest-neighbour distance of their hexagonal lattice is about 5 Å.

These later phases exhibit the above-mentioned modulated textures, containing disclinations and walls [80]. Many other modulated structures were observed experimentally in thin free-standing smectic films (e.g. [81–83]). Such structures will be considered in more detail in Chapter 7, in connection with the analysis of topological defects in FSLC films.

Characteristically for two-dimensional systems, Landau–Peierls divergence of fluctuations is observed experimentally at the surfaces of free-standing DOBAMBC films (by means of intensity-fluctuation spectroscopy and depolarised reflection microscopy) [84, 85]. Note that the characteristic anomaly for xy-systems in the

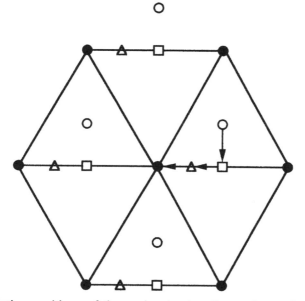

Fig. 5.9 Relative positions of the molecules in adjacent layers in the different crystalline B phases. The solid dots represent the reference layer positions. The open symbols show the layer positions in the next layer above for lattices with a single-layer unit cell. For lattices with a two-layer unit cell, the position of the layer below is the same as that for the layer above. The open circles represent both the hexagonal ABC structure (one-layer unit cell) and the hexagonal AB AB structure (two-layer unit cell). The squares indicate the orthorhombic structure positions. The triangles show both the monoclinic (one-layer unit cell) and the orthorhombic-a (two-layer unit cell) structures. In the hexagonal AAA phase the molecular positions in the adjacent layers are the same as those in the reference layer (reproduced with permission from Reference [79]. Copyright 1987 by the American Physical Society)

temperature behaviour of the elastic constants, described by Eq. (5.31), has not, however, been directly detected experimentally.

Some quantitative data on two-dimensional BO ordering were obtained for free-standing films of structured smectics with hexatic in-plane molecular organisation by means of light scattering [86–88], X-ray [89, 90] and electron diffraction [91, 92] (see also Chapter 3). It has been shown, for example, that for free-standing films of 4-(2′-methylbutyl)phenyl-4′-(octyloxy)-(1,1′)-biphenyl-4-carboxylate (8OSI) in the smectic I phase, the BO order parameter C_6 ($n = 1$), calculated from formula (3.15), has values of the order of 0.1–0.9 (the thinner the film, the smaller is C_{6n}), and that it goes smoothly to zero in the vicinity of the transition temperature to the smectic C phase [89, 90]. Moreover, for different harmonics of C_{6n}, the scaling law is found to be well described by Eq. (5.28) [89, 90].

Note, finally, that phase transitions in thermotropic FSLC films can be induced not only by temperature changes. Thus phase transitions under the action of an external electric field are also predicted theoretically for monolayer free-standing films [52]. An interesting example of first-order smectic A–smectic C* phase transfer, initiated by an increase in the film surface tension (by means of film area variation), is reported in Reference [93].

REFERENCES

1. Sonin A.A., Yethiraj A., Bechhoefer J., Frisken B., 1995, *Phys. Rev. E*, **52**, 6260.
2. Meyerhoffer D., Sussman A., Williams R., 1972, *J. Appl. Phys.*, **43**, 3685.
3. Faetti S., Fronzoni L., 1978, *Solid State Commun.*, **25**, 1087.
4. Sonin A.A., 1995, *The Surface Physics of Liquid Crystals*, OPA-Gordon and Breach, Amsterdam.
5. Mada H., 1979, *Mol. Cryst. Liq. Cryst.*, **51**, 43.
6. Mada H., 1979, *Mol. Cryst. Liq. Cryst.*, **53**, 127.
7. Parsons J.D., 1978, *Phys. Rev. Lett.*, **41**, 877.
8. Barbero G., Dozov I., Palierne J.F., Durand G., 1986, *Phys. Rev. Lett.*, **56**, 2056.
9. Blinov L.M., Chigrinov V.G., 1994, *Electrooptic Effects in Liquid Crystal Materials*, Springer, New York.
10. Parsons J.D., 1976, *J. Phys. (France)*, **37**, 1187.
11. Barbero G., Durand G., 1990, *J. Appl. Phys.*, **67**, 2678.
12. Barbero G., Durand G., 1990, *J. Phys. (France)*, **51**, 281.
13. Russel W.B., Saville D.A., Schowalter W.R., 1991, in *Colloidal Dispersions*, Cambridge University Press, Cambridge, UK, p. 102.
14. Yablonsky S.V., Blinov L.M., Pikin S.A., 1985, *Mol. Cryst. Liq. Cryst.*, **127**, 381.
15. Guyot-Sionnest P., Hsiung H., Shen Y.R., 1986, *Phys. Rev. Lett.*, **57**, 2963.
16. Chuvyrov A.N., Lachinov A.N., 1978, *Zh. Eksp. Teor. Fiz.*, **74**, 1431.
17. Lachinov A.N., Chuvyrov A.N., Sonin A.S., 1982, *Fiz. Tverd. Tela (Leningrad)*, **24**, 255.
18. Alexe-Ionescu A.L., Barbero G., Petrov A.G., 1993, *Phys. Rev. E*, **48**, R1631.
19. Durand G., 1990, *Physica A*, **163**, 94.
20. Faetti S., Palleschi V., 1984, *Phys. Rev. A*, **30**, 3241.
21. Chuvyrov A.N., 1980, *Kristallografiya*, **25**, 326.
22. Barbero G., Chuvyrov A.N., Kaniadakis G., Miraldi E., Rastello M.L., 1993, *J. Phys. II*, **3**, 165.

23. Dubois-Violette E., de Gennes P.G., 1975, *J. Phys. (France) Lett.*, **36**, L-255.
24. Ryschenkow G., Kléman M., 1976, *J. Chem. Phys.*, **64**, 404.
25. Blinov L.M., Davydova N.N., Sonin A.A., Yudin S.G., 1984, *Kristallografiya*, **29**, 537.
26. Bahr Ch., Booth C.J., Fliegner D., Goodby J.W., 1996, *Europhys. Lett.*, **34**, 507.
27. Sheng P., 1976, *Phys. Rev. Lett.*, **37**, 1059.
28. Sheng P., 1982, *Phys. Rev. A*, **26**, 1610.
29. Sonin A.A., 1988, *Poverkhnost*, No. 11, 42.
30. Pavel J., 1984, *J. Phys. (France)*, **45**, 137.
31. Cahn J.W., 1977, *J. Chem. Phys.*, **66**, 3667.
32. Pandit R., Wortis M., 1982, *Phys. Rev. B*, **25**, 322.
33. Brézin E., Halperin B.I., Leibler S., 1983, *J. Phys. (France)*, **44**, 775.
34. Hauge E.H., 1986, *Phys. Rev. B*, **33**, 3322.
35. de Gennes P.G., 1974, *The Physics of Liquid Crystals*, Clarendon Press, Oxford, UK.
36. de Gennes P.G., Prost J., 1993, *The Physics of Liquid Crystals*, 2nd edn., Clarendon Press, Oxford, UK.
37. Landau L.D., Lifshitz E.M., Pitaevskii L.P., 1980, *Statistical Physics*, 3rd edn., Pergamon Press, New York.
38. Peierls R.E., 1974, *Helv. Phys. Acta., Suppl.*, **11**, 81.
39. Young C.Y., Pindak R., Clark N.A., Meyer R.B., 1978, *Phys. Rev. Lett.*, **40**, 773.
40. Rosenblatt Ch., Pindak R., Clark N.A., Meyer R.B., 1979, *Phys. Rev. Lett.*, **42**, 1220.
41. Rosenblatt Ch., Meyer R.B., Pindak R., Clark N.A., 1980, *Phys. Rev. A*, **21**, 140.
42. Aharony A., Birgeneau R.J., Brock J.D., Litster J.D., 1986, *Phys. Rev. Lett.*, **57**, 1012.
43. Kosterlitz J.M., Thouless D.J., 1972, *J. Phys. C*, **5**, L124.
44. Kosterlitz J.M., Thouless D.J., 1973, *J. Phys. C*, **6**, 1181.
45. Nelson D.R., Kosterlitz J.M., 1977, *Phys. Rev. Lett.*, **39**, 1201.
46. Amador S.M., Pershan P.S., 1990, *Phys. Rev. A*, **41**, 4326.
47. Halperin B.I., Nelson D.R., 1978, *Phys. Rev. Lett.*, **41**, 121.
48. Young A.P., 1979, *Phys. Rev. B*, **19**, 1855.
49. Langer St., Sethna J.P., 1986, *Phys. Rev. A*, **34**, 5035.
50. Hinshaw G.A., Jr., Petschek R.G., Pelcovits R.A., 1988, *Phys. Rev. Lett.*, **60**, 1864.
51. Hinshaw G.A., Jr., Petschek R.G., 1989, *Phys. Rev. A*, **39**, 5914.
52. Chen C.-M., MacKintosh F.C., 1995, *Europhys. Lett.*, **30**, 215.
53. Ohyama T., Jacobs A.E., Mukamel D., 1996, *Phys. Rev. E*, **53**, 2595.
54. Strandburg K.J., 1988, *Rev. Mod. Phys.*, **60**, 161.
55. Pershan P.S., 1988, *Structure of Liquid Crystal Phases*, World Scientific, Singapore.
56. Geer R., Stoebe T., Huang C.C., 1993, *Phys. Rev. E*, **48**, 408.
57. Bahr Ch., 1994, *Int. J. Mod. Phys. B*, **8**, 3051.
58. Stoebe T., Huang C.C., 1995, *Int. J. Mod. Phys. B*, **9**, 2285.
59. Bahr Ch., Fliegner D., 1992, *Phys. Rev. A*, **46**, 7657.
60. Heinekamp S., Pelcovits R.A., Fontes E., Chen E.Y. *et al.*, 1984, *Phys. Rev. Lett.*, **52**, 1017.
61. Kraus I., Pieranski P., Demikhov E., Stegemeyer H., Goodby J., 1993, *Phys. Rev. E*, **48**, 1916.
62. Heinekamp S.W., Pelcovits R.A., 1985, *Phys. Rev. A*, **32**, 2506.
63. Davey S.C., Budai J., Goodby J.W., Pindak R., Moncton D.E., 1984, *Phys. Rev. Lett.*, **53**, 2129.
64. Bahr Ch., Booth C.J., Fliegner D., Goodby J.W., 1996, *Phys. Rev. Lett.*, **77**, 1083.
65. Sirota E.B., Pershan P.S., Amador S., Sorensen L.B., 1987, *Phys. Rev. A*, **35**, 2283.
66. Amador S., Pershan P.S., Stragier H., Swanson B.D. *et al.*, 1989, *Phys. Rev. A*, **39**, 2703.
67. Tweet D.J., Holyst R., Swanson B.D., Stragier H., Sorensen L.B., 1990, *Phys. Rev. Lett.*, **65**, 2157.
68. Stoebe T., Huang C.C., 1994, *Phys. Rev. E*, **50**, R32.

69. Stoebe T., Ho J.T., Huang C.C., 1994, *Int. J. Thermophys.*, **15**, 1189.
70. Jin A.J., Stoebe T., Huang C.C., 1994, *Phys. Rev. E*, **49**, R4791.
71. Chao C.Y., Chou C.F., Ho J.T., Hari S.W., *et al.*, 1996, *Phys. Rev. Lett.*, **77**, 2750.
72. Chou C.F., Jin A.J., Chao C.Y., Hui S.W. *et al.*, 1997, *Phys. Rev. E*, **55**, R6337.
73. Ocko B.M., Braslau A., Pershan P.S., Als-Nielsen J., Deutsch M., 1986, *Phys. Rev. Lett.*, **57**, 94.
74. Stoebe T., Mach P., Huang C.C., 1994, *Phys. Rev. Lett.*, **73**, 1384.
75. Mirantsev L.V., 1995, *Phys. Lett. A*, **205**, 412.
76. Martinez Y., Somoza A.M., Mederos L., Sullivan D.E., 1996, *Phys. Rev. E*, **53**, 2466.
77. Collett J., Pershan P.S., Sirota E., Sorensen L.B., 1984, *Phys. Rev. Lett.*, **52**, 356.
78. Sirota E., Pershan P.S., Sorensen L.B., Collett J., 1985, *Phys. Rev. Lett.*, **55**, 2039.
79. Sirota E., Pershan P.S., Sorensen L.B., Collett J., 1987, *Phys. Rev. A*, **36**, 2890.
80. Sirota E., Pershan P.S., Deutsch M., 1987, *Phys. Rev. A*, **36**, 2902.
81. Demikhov E.I., 1995, *Pis'ma Zh. Eksp. Teor. Fiz.*, **61**, 951.
82. Demikhov E.I., Stegemeyer H., 1995, *Liq. Cryst.*, **18**, 37.
83. Pang J., Clark N.A., 1994, *Phys. Rev. Lett.*, **73**, 2332. 71.
84. Van Winkle D.H., Clark N., 1984, *Phys. Rev. Lett.*, **53**, 1157.
85. Van Winkle D.H., Clark N., 1988, *Phys. Rev. A*, **38**, 1573.
86. Dierker S.B., Pindak R., 1987, *Phys. Rev. Lett.*, **59**, 1002.
87. Sprunt S., Litster J.D., 1987, *Phys. Rev. Lett.*, **59**, 2682.
88. Sprunt S., Spector M.S., Litster J.D., 1992, *Phys. Rev. A*, **45**, 7355.
89. Brock J.D., Aharony A., Birgeneau R.J., Evans-Lutterodt K.W. *et al.*, 1986, *Phys. Rev. Lett.*, **57**, 98.
90. Brock J.D., Noh D.Y., McClain B.R., Litster J.D. *et al.*, 1989, *Z. Phys. B*, **74**, 197.
91. Cheng M., Ho J.T., Hui S.W., Pindak R., 1987, *Phys. Rev. Lett.*, **59**, 1112.
92. Cheng M., Ho J.T., Hui S.W., Pindak R., 1988, *Phys. Rev. Lett.*, **61**, 550.
93. Kraus I., Pieranski P., Demikhov E., 1994, *J. Phys.: Condens. Matter*, **6**, A415.

6

Field-Induced Effects

In this chapter we will analyse the characteristics of some physical effects induced in FSLC films by external (electric or magnetic) fields. These are EMO (electro-magneto-optical) phenomena: the Fredericks transition, the flexoelectric effect (see also Chapters 1 and 5) and the EHD (electrohydrodynamic) instabilities (see also Chapter 1) in free-standing thermotropic nematic films. The characteristic features of these phenomena in freely suspended films are connected mainly with the specific experimental geometry (e.g. the presence of a thicker liquid meniscus at the film border) and with peculiar (degenerated) boundary conditions, allowing the two-dimensional motion of the fluid. In conclusion, EHD instabilities in smectic A, C, C* free-standing films and some other examples of the electric field-induced effects in freely suspended smectic C* films will be considered.

6.1 THE FREDERICKS TRANSITION

The Fredericks effect in the freely suspended, spontaneously homeotropically oriented, nematic 5CB films (see Section 5.1) has been studied experimentally† in the film-supporting frame, shown in Fig. 3.2. The Fredericks transition is observed both in electric (a.c. voltage is applied to the conducting wires 1 and 2) and in magnetic (the film is placed between the poles of an electromagnet) fields. In both cases, the character of the director distortion is very similar. The Fredericks transition starts from the meniscus areas, where the film thickness is greater than in the flat portion of the film (see Eq. 1.17), Figs. 6.1(a) and 6.1(b). When the applied external field increases, the two director deformation modes (with inverse signs of the director tilt) propagate from both the meniscuses in the direction of the film centre. For $E \gg E_F$, they come close together, and a defect wall is formed between them, in the centre of the film (Fig. 6.1(c)). As has already been mentioned

---

† A.A. Sonin, unpublished work.

(see Chapter 1), such walls may also be observed in the course of the Fredericks transition in nematic layers confined between solid substrates (e.g. [1, 2]).

Note that there is no visible hydrodynamic motion of the fluid in the course of the Fredericks transition: small dust particles, placed at the upper film interface, do not move.

6.2 FLEXOELECTRIC EFFECT

The flexoelectric effect may also be observed in the experimental geometry of Fig. 3.2 for homeotropic free-standing films prepared from nematics with a negative dielectric anisotropy (MBBA for instance). The flexo-deformation should be greater in the thicker meniscus regions of the film (according to the formula 1.18). The

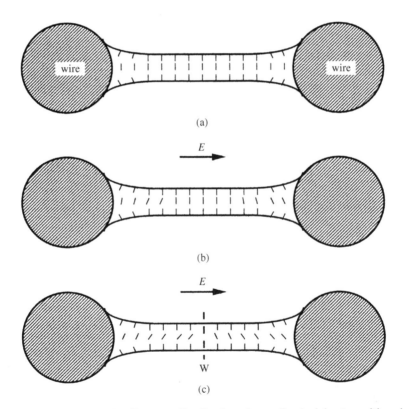

Fig. 6.1 The schematic director distribution for a Fredericks transition in an initially homeotropic freely suspended 5CB film in an a.c. electric field (A.A. Sonin, unpublished work): (a) $E = 0$; (b) $E \geqslant E_F$; (c) $E \gg E_F$; W is the topological wall

developing EHD instability (see the next paragraph), however, usually screens the flexo-effect.†

6.3 ELECTROHYDRODYNAMIC INSTABILITIES

The EHD instabilities of liquid crystals are usually investigated in liquid layers sandwiched between two parallel glass plates. The latter affects strongly the EHD motion by fixing the boundary conditions and the thickness of the layer. Thus, in this case, a two-dimensional flow is not allowed.

In the absence of the glass plates (freely suspended films), the situation is quite different. The local thickness of the film is now a variable parameter coupled to the hydrodynamic flow; and a two-dimensional liquid motion can occur.

The two-dimensional EHD motion was observed in free-standing films made of certain thermotropic mesophases: nematic [1–5], smectic A [6–8] and smectic C and C* [9–11]. The films were prepared on rectangular supporting frames of the type shown in Fig. 3.2. The d.c. or a.c. electric fields were applied to the conducting wires 1 and 2 of such frames. The distance a between these wires was varied within the limits 0.1–several mm. The film thicknesses, h, were changed between several μm and several tens of μm (for nematic films), between 0.06 and 0.8 μm (for smectic A films) and between 0.1 and 3 μm (for smectic C and C* films).

The EHD instability observations were realised by means of polarisation-optical microscopy in both transmitted and reflected light. In some cases, for better visualisation of the flow patterns, small dust or smoke particles were deposited at the film interfaces.

6.3.1 NEMATIC FILMS

The two following different modes of EHD instabilities have been observed experimentally (by means of optical microscopy in transmitted and reflected light) in free-standing nematic films [1–5]:

(i) the domain mode (DM), which consists of a series of adjacent elongated domains with a constant spatial period. This mode occurs if h exceeds a critical value $h_c \approx 7$ μm and appears only in the anisotropic state of nematics (i.e. for $T < T_{NI}$, where T_{NI} is the nematic–isotropic phase transition temperature);

(ii) the vortex mode (VM), which consists of a series of adjacent vortices with a constant spatial period, and which appears in thin nematic films ($h < h_c$). This mode occurs both in the anisotropic ($T < T_{NI}$) and isotropic ($T > T_{NI}$) states of nematics. The first observation of such instability was made many years ago for some isotropic fluids [12].

---

† A.A.Sonin, unpublished work.

We will consider in more detail the two just-mentioned types of the EHD instabilities. Our analysis of the DM will be based mostly on the experimental and theoretical results described in Reference [4]. The most distinct domain EHD patterns are obtained for the homeotropically oriented MBBA (negative dielectric anisotropy) free-standing films. The uniform homeotropic alignment of the director is achieved by choosing the appropriate film thickness (see also Chapter 5), by the appropriate choice of the ratio between the film length and width (b/a), see Fig. 3.2.

The DM occurs in both d.c. and a.c. electric fields, if the electric voltage exceeds a threshold value U_{cd}. The regular domain pattern disappears if the frequency f of the applied a.c. electric field becomes greater than the relaxation frequency, f_r, of the ionic electric charges in the nematic film. The value of f_r is of the order of

$$f_r = 2(\sigma/\varepsilon), \tag{6.1}$$

where σ and ε are functions of the electric conductivity coefficient and the permittivity coefficient, respectively. For MBBA films, studied in [2–4], $f_r \approx 300\,\text{Hz}$ at $T = 25\,^\circ\text{C}$.

Depending on the frequency of the applied electric field, two different domain patterns are observed: the low frequency texture (LFT, $f < 100\,\text{Hz}$) and the high frequency texture (HFT, $f > 100\,\text{Hz}$). They are shown in Figs. 6.2(a) and 6.2(b), respectively.

The LFT appears over the whole film (as soon as the applied electric voltage is greater than U_{cd}), while the HFT starts from the conducting wires. The d.c. domains are similar to the low frequency ones, but they are less regular. By increasing f one can observe a progressive transition from the LFT to the HFT without lack of continuity. Moreover, the threshold voltages of the LFT and HFT show the same qualitative dependence of all physical parameters of the film. This means that, probably, the same physical process is involved in the generation of these two different patterns.

The period λ_d of the domain texture (see Fig. 6.2) is weakly dependent on the voltage amplitude, temperature and electric field frequency, while it strongly depends on the film thickness (as for the Williams domains described in Chapter 1):

$$\lambda_d = (6 \pm 1)h. \tag{6.2}$$

The hydrodynamic flow inside the domain structure considered is detected by observing the motion of small particles suspended at the film interface. Typical values of the hydrodynamic velocity, v, are of the order of $10^{-2}\,\text{cm/s}$. The flow lines in the film plane, (x, y), show the same qualitative behaviour as the director field represented in Fig. 6.2.

For the DM threshold voltage, the following empirical law is found:

$$U_{cd} = \beta a/h, \tag{6.3}$$

where β is a coefficient dependent on frequency and temperature.

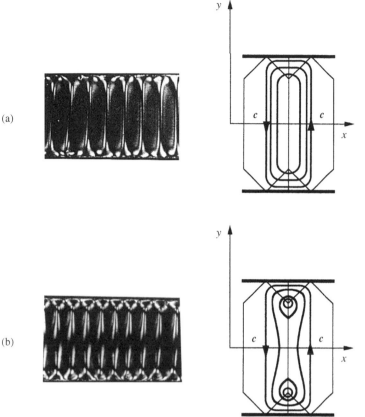

Fig. 6.2 Views (in crossed microscope, of the angle between the polariser and the x-axis is 39°) and the corresponding distributions of the two-dimensional director, **c** (inside a single domain) for the domain mode Nicols EHD instability in MBBA free-standing film (reprinted with permission from Reference [4]. Copyright 1983 American Institute of Physics): (a) low frequency domain pattern; (b) high frequency domain pattern

The value of β shows a smooth increase with the frequency if $f \ll f_r$, while β increases greatly if $f \approx f_r$. This behaviour is also very similar to that observed in the Williams domain mode in nematics confined between solid substrates (see Chapter 1 and Refs. [13, 14]). For $f = 20$ Hz and $T = 25\,°C$, it is found that $\beta = (1.8 \pm 0.3)$ V.

When the voltage applied to the electrode wires grows considerably, the domain pattern becomes less stable and regular. For example, if U is raised up to three times that of U_{cd}, the domain pattern changes into a dynamic scattering regime.

The domain pattern character does not alter if the metallic electrode wires are covered by an insulating paint. This means that the ion injection from the electrodes does not play an important role in the DM of the EHD instability, but that this phenomenon is most probably described by the Carr–Helfrich mechanism (see Chapter 1). The distribution of the two-dimensional charges produced in our case by the Carr–Helfrich effect is shown in Fig. 6.3.

Let us describe a simple phenomenological model of the DM EHD instability in freely suspended nematic films. A schematic geometry of the problem considered is shown in Fig. 6.4.

To obtain analytical results, one needs to make the following assumptions:

1. the initial director orientation in the film is homeotropic;
2. the applied electric field is d.c. (in this case, the electric field inside the film will be homogeneous, except the regions neighbouring the conducting wires);

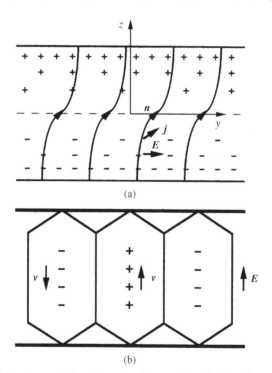

(a)

(b)

Fig. 6.3 Schematic view of the electric charge distribution in a nematic free-standing film generated by the Carr–Helfrich effect (reprinted with permission from Reference [4]. Copyright 1983 American Institute of Physics). (a) Cross-section of the film: n indicates the local orientation of the director, j is the electric current density, E is the electric field. Because of the anisotropy of the electric conductivity, the distortion of n generates the electric charge distribution shown in this Figure. (b) Top view of the film with a schematic arrangement of the induced electric charge distribution near the top interface: electric charges with opposite signs lie near the bottom surface of the film; v indicates the hydro-dynamic velocity near the top film interface

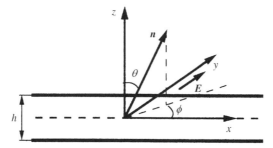

Fig. 6.4 The geometry of the considered domain mode EHD instability problem (reprinted with permission from Reference [4]. Copyright 1983 American Institute of Physics)

3. the film thickness is homogeneous (this is true in the central portion of the film, while near the wires, h increases abruptly);
4. the director anchoring at the film interfaces is strong;
5. the hydrodynamic flow does not modify the planarity of the film interfaces;
6. the hydrodynamic velocity is directed along the y axis only ($v = v_y$).

   The threshold of the DM can be calculated by using the method of linear analysis of the instability [15]: one must study the stability of the system under small two-dimensional fluctuations of the director tilt angle, $\delta\theta$, the flow velocity, δv_y, and the electric potential, $\delta\Phi$. Therefore, one should look for the solutions for θ, v and Φ in the following form:

$$\theta = \theta_0 + \delta\theta = \theta_0 + \theta(z)\,\exp(ik_dx)\,\exp(c_dt), \tag{6.4}$$

$$v = \delta v_y = v(z)\,\exp(ik_dx)\,\exp(c_dt), \tag{6.5}$$

$$\Phi = \Phi_0 + \delta\Phi = \Phi_0 + \Phi(z)\,\exp(ik_dx)\,\exp(c_dt). \tag{6.6}$$

Here θ_0, v_0 and Φ_0 are the stationary values of, respectively, the director tilt angle, hydrodynamic velocity and electric potential; $k_d = 2\pi/\lambda_d$ is the DM instability wave-vector; and c_d is a constant.

   At the instability threshold the fluctuations neither decay nor grow, then $c_d = 0$.

   One can write the following (linear with respect to θ, v and Φ) Maxwell's and Leslie–Ericksen's equations:

$$\frac{\partial^2\Phi}{\partial z^2} - \frac{\Delta\sigma}{\sigma_\parallel}E\frac{\partial\theta}{\partial z} - \frac{\sigma_\perp}{\sigma_\parallel}k_d^2\Phi = 0, \tag{6.7}$$

$$\frac{\partial^2\Phi}{\partial z^2} - k_d^2\Phi = \frac{4\pi\rho\Delta\sigma}{\varepsilon_\perp\sigma_\parallel - \varepsilon_\parallel\sigma_\perp}, \tag{6.8}$$

$$\rho E = \eta_a k_d^2 v - \eta_c \frac{\partial^2 v}{\partial z^2}, \tag{6.9}$$

$$-\alpha_2 \frac{\partial v}{\partial z} - K_{33} \frac{\partial^2 \theta}{\partial z^2} + K_{22} k_d^2 \theta - \frac{\Delta \varepsilon}{4\pi} \left(E^2 \theta - E \frac{\partial \Phi}{\partial z} \right) = 0. \tag{6.10}$$

Here σ and ε are the electric conductivity and the electric permittivity, respectively; the indices '$\|$' and '$\perp$' refer to the direction of measurements parallel and perpendicular to the director, respectively; $\Delta\sigma = \sigma_\| - \sigma_\perp$, $\Delta\varepsilon = \varepsilon_\| - \varepsilon_\perp$; η_a, η_c, α_2 are the viscosity coefficients; K_{22} and K_{33} are the twist and bend elastic constants, respectively; $E = U/a$ is the average electric field; and ρ is the bulk space charge density.

The following boundary conditions must be satisfied at the free surfaces ($z = \pm h/2$):

$$\theta = 0, \tag{6.11}$$

$$\frac{\partial v}{\partial z} = 0, \tag{6.12}$$

$$\frac{\partial \Phi}{\partial z} = 0. \tag{6.13}$$

Equations (6.7)–(6.10) have the following solutions of the form (6.4)–(6.6):

$$\theta = \theta_0 \cos(N\pi z/h), \tag{6.14}$$

$$v = v_0 \sin(N\pi z/h), \tag{6.15}$$

$$\Phi = \Phi_0 \sin(N\pi z/h), \tag{6.16}$$

where N is an odd integer.

The threshold voltage of the DM will then be given by Eq. (6.3) with

$$\beta = \left\{ \frac{4\pi^3(K_{33} + K_{22}s)}{\left(\frac{1+s}{\sigma_\| + \sigma_\perp s}\right)\left[\Delta\varepsilon\sigma_\perp - \frac{\alpha_2(\varepsilon_\perp\sigma_\| - \varepsilon_\|\sigma_\perp)}{\eta_c + \eta_a s}\right]} \right\}^{1/2}, \tag{6.17}$$

where $s = (k_d^2 h^2)/\pi^2$.

It is clear from Eq. (6.17) that for an isotropic fluid ($\Delta\varepsilon = 0$ and $\Delta\sigma = 0$) $U_{cd} \to \infty$, i.e. the DM of the EHD instability does not exist.

According to the just-described experimental data $k_d \propto h^{-1}$, so $s \propto (\pi)^{-2}$. By substituting the experimental values of the MBBA material parameters (e.g. [4, 13, 14]): $\varepsilon_\perp = 5.4$, $\varepsilon_\| = 4.7$; $\alpha_2 = -0.77$ P, $\eta_2 = 0.24$ P, $\eta_c = 0.41$ P; $K_{22} = 4 \times 10^{-7}$ dyne, $K_{33} = 7 \times 10^{-7}$ dyne, one finds $\beta = 1.75$ V. The obtained estimation are in a good agreement with the experimental data (see above).

Note that similar results can be obtained in the case of low-frequency a.c. electric fields. In fact, such fields are almost uniform inside the film. Thus for $f \ll f_r$, one should expect that the mean square value of the threshold voltage is given by

$$U_{cd}(f) = U_{cd} = \beta(a/h) \approx (1.75a/h)\,\mathrm{V}, \qquad (6.18)$$

where U_{cd} is the threshold voltage in the d.c. case.

Eq. (6.18) is also in good agreement with the experimental results.

When high-frequency electric fields are used, analytic relations cannot be obtained, due to the strong and complex dependence of the electric field on the y co-ordinate. This implies that a two-dimensional analysis cannot be used. However, by analogy with the EHD instabilities in nematic bounded by solid substrates in the conductivity regime (see Chapter 1), one should expect that the threshold voltage greatly increases when the electric field frequency approaches f_r. Indeed, this behaviour has been observed experimentally.

Now we will discuss experimental and theoretical results of [5], concerning the VM (vortex mode). The VM is observed in MBBA free-standing films in both d.c. and a.c. electric fields. The most distinct vortex patterns are obtained for the uniform homeotropic alignment of the director. The vortices manifest themselves as a series

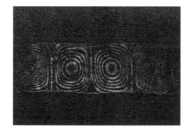

(a)

(b)

Fig. 6.5 The vortex mode EHD of instability in a freely suspended MBBA film (reprinted with permission from Reference [5]. Copyright 1983 American Institute of Physics): (a) Newton interference fringes in the presence of the vortex flow (observed in a reflection microscope); (b) schematic hydrodynamic flow lines in the film plane. The lines are inclined possibly due to the inhomogeneity of the film thickness. Photograph courtesy of Dr S. Faetti

of adjacent, nearly elliptical, interference fringes [Fig. 6.5(a)]. These patterns are due to the periodic inhomogeneities of the film thickness, modulated by the liquid motion. It is found that the local film thickness has a maximum value at the boundary between adjacent vortices and a minimum value at the centre of every vortex.

Quite often, in the case of the d.c. electric field, the oval interference patterns associated with the vortex flow are strongly tilted. Fig. 6.5(b) shows schematically the hydrodynamic flow lines for such a vortex pattern. This tilt is most probably due to the coupling between the hydrodynamic vortex motion and the non-uniform thickness profile of the film.

Due to the presence of two free surfaces, the liquid motion in the vortices is characterised by high velocities (of the order of 10 cm/s). The spatial period of the vortex pattern, λ_v, [Fig. 6.5(b)] is found to be proportional to the distance a between the conducting wires (see Fig. 3.2)

$$\lambda_v = \alpha a. \tag{6.19}$$

Here $\alpha = 1.4 \pm 0.3$.

The vortex instability has a threshold character. The threshold voltage U_{cv} is found to be proportional to the average film thickness $\langle h \rangle$; ($\langle h \rangle = (V_0/ab)$, where V_0 is the film volume, and a and b are the dimensions of the rectangular film-supporting frame). U_{cv} also depends upon the electric conductivity of the film ($U_{cv} \propto \sqrt{I}$, where I is the electric current flowing across the film).

In the case of the a.c. electric field, the fluid motion inside the vortices inverts its direction when the polarity of the electric voltage is reversed, and thus the vortex pattern oscillates with the same frequency as the a.c. electric field. The EHD instability in this case also has a threshold character. The threshold amplitude U_{cv}' is growing with the increasing electric field frequency. At low frequencies ($f < 10$ Hz) $U_{cv}' \approx U_{cv}$ (U_{cv} is the threshold for the d.c. electric field). Furthermore, the a.c. vortices have the same geometrical dimensions as the d.c. vortices. These observations suggest that the a.c. and d.c. vortex patterns are generated by the same physical mechanism.

The injection of electric charges from electrodes does not play a role in the VM EHD instability, as is the case for the DM (see above). Indeed, the same vortex patterns have also been observed with the wires covered with insulating paint.

In contrast, the presence of electric charges at the film interfaces plays a crucial role in the formation of the vortex patterns. This fact was proved by the following experiment [3]. A d.c. voltage, U, was applied between two conducting wires to generate an electric field, E, in the nematic film. At the same time, a d.c. voltage, U^*, was applied between two electrodes external to the film (Fig. 6.6). Because of the electric conductivity of MBBA, some electric ionic charges accumulate at the free surfaces of the film in such a way as to screen the electric field generated by the external electrodes (see Fig. 6.6). Then, in the stationary state, the external electric field cannot penetrate the film. Therefore the external voltage, U^*, contributes towards modifying the charge distribution at the free surfaces, but does not alter the electric field inside the film. If the VM is due to bulk space charges, the threshold voltage, U_{cv}, would not depend on the external voltage, U^*. On the contrary, it has been found that U^* affects considerably the U_{cv} values. Thus the presence of the surface electric charges plays an important role in the formation of the vortex patterns.

Consider now a simple theoretical model of the VM of the EHD instability. The vortex instability is interpreted in terms of the interaction between the electric field, E, and the electric charges lying at the film's free surfaces. These electric charges accumulate at the liquid–air interfaces because of the discontinuity of the electric conductivity. The stationary distribution of the charges is reached after some characteristic time, τ, after the switching on the electric field, when the component of E perpendicular to the film interfaces vanishes (inside the film). The mechanism which generates the instability is schematically shown in Fig. 6.7.

The vortex instability threshold voltage, U_{cv}, can be obtained (as in the case of the DM) by solving the Navier–Stokes equations of the hydrodynamic motion and the Maxwell equations of the electric field. The characteristics of the considered surface charge model (SCM) are related to the relevant role that is played by the boundary conditions. This model is similar to the one characterising the well-known thermodynamic Marangoni's instability [16] in fluids.

Because of the complexity of the mathematical problem related to the SCM, we will make some simplifying assumptions.

1. Since, experimentally, the vortex patterns are observed both in anisotropic and isotropic fluids, we will neglect the anisotropy of the liquid crystal. So, we will

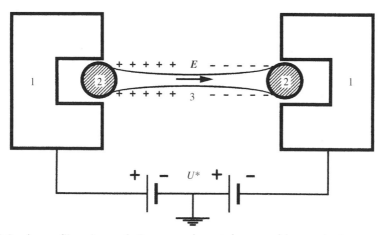

Fig. 6.6 A profile view of the experimental assembly made to prove the important role of surface electric charges in the formation of the vortex patterns: (reproduced with permission of EDP Sciences from Reference [3]): (1) external electrodes; (2) metal wires; (3) a film

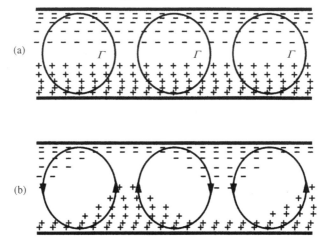

Fig. 6.7 Mechanism of the vortex instability in a free-standing liquid film (reprinted with permission from Reference [5]. Copyright 1983 by the American Institute of Physics): (a) Top view of the film showing the stationary surface charge distribution below the threshold of the VM. In this regime the surface charge distribution is uniform along the *x* axis parallel to the conducting wires, therefore the average value of the surface electric force along the closed flow lines, Γ, is zero. (b) Top view of the surface charge distribution when a vortex fluctuation occurs. The hydrodynamic flow drags electric charges and generates a periodic surface charge density. The mean value of the surface electric force along a closed flow line is now different from zero and is directed in such a way to increase the hydrodynamic motion. If the electric voltage is greater than the threshold value, U_{cv}, then the surface electric force will exceed the viscous one, and a macroscopic vortex pattern will occur

consider the isotropic, incompressible freely suspended liquid film having two plane free surfaces parallel to the (x, y) plane of a Cartesian orthogonal co-ordinate system (see Fig. 6.4).

2. The diffusion of charges and the variation of the surface profile due to the action of the electric field and to the hydrodynamic motion will also be neglected.

The boundary condition responsible for the instability is given by the balance of charges at the film interfaces:

$$\frac{\partial q}{\partial t} = \sigma E \cdot L - \nabla_2 \cdot (j + qv). \tag{6.20}$$

Here q is the surface charge density per unit area, σ is the electric conductivity of the film medium, E is the electric field inside the film just below the surface charge, L is the unit vector normal to the free surface and outward directed, ∇_2 is the two-dimensional Laplacian in the (x, y) horizontal plane, $\boldsymbol{v}$ is the hydrodynamic velocity near the surface and $\boldsymbol{j}$ is the electric current per unit length, defined as

$$\boldsymbol{j} = \int_{\bar{z}-L_s}^{\bar{z}} \sigma \boldsymbol{E} \, \mathrm{d}z, \tag{6.21}$$

where $\bar{z}(> 0)$ is the co-ordinate of the free surface and L_s is the thickness of the interface layer with the surface charge distribution, which is of the order of the Debye length, L_D ($\approx 0.1 \, \mu\mathrm{m}$ for the MBBA film used in Reference [5]).

Since L_s is very small with respect to the experimental values of the film thickness ($h \approx 10 \, \mu\mathrm{m}$), the contribution of $\boldsymbol{j}$ in Eq. (6.20) can be neglected. With this assumption and by using the incompressibility equation ($\nabla \boldsymbol{v} = 0$, i.e. $\partial v_z / \partial z = -\nabla_2 \boldsymbol{v}$), Eq. (6.20) can be rewritten as

$$\frac{\partial q}{\partial t} = \sigma \boldsymbol{E} \cdot \boldsymbol{L} + q \frac{\partial v_z}{\partial z} - \boldsymbol{v} \cdot \nabla_2 q. \tag{6.22}$$

The other electric and hydrodynamic boundary conditions at the film interfaces are

$$q = \frac{1}{4\pi} (\varepsilon \nabla \Phi_i - \nabla \Phi_e) \cdot \boldsymbol{L}, \tag{6.23}$$

$$\Phi_e = \Phi_i, \tag{6.24}$$

$$q \frac{\partial \Phi_i}{\partial x} = -\eta \frac{\partial v_x}{\partial L}, \tag{6.25}$$

$$q \frac{\partial \Phi_i}{\partial y} = -\eta \frac{\partial v_y}{\partial L}, \tag{6.26}$$

where $\partial / \partial L$ indicates the derivative along the $\boldsymbol{L}$ direction; ε and η are the dielectric constant and the viscosity coefficient, respectively; and Φ_i and Φ_e are the electric potentials internal and external to the conducting fluid, respectively.

Since we neglect the inhomogeneity of the film thickness profile, we have omitted the boundary condition for the vertical stresses due to the pressure, the electric field, and the surface tension γ. This assumption corresponds to the limiting case $\gamma \rightarrow 0$.

The boundary conditions at the conducting wires are

$$\Phi = \pm \frac{U}{2}, \tag{6.27}$$

where U is the electric voltage applied to the wires.

The source of the vortex instability is contained in the boundary conditions (6.22), (6.25) and (6.26) which couple the hydrodynamic flow to the surface charges.

Taking into account the experimental fact that there is no considerable injection of charges from electrodes and from the diffusion layer of bulk space charges close to the electrodes, we can write the following electrohydrodynamic equations for the bulk of the film:

$$\nabla^2 \Phi = 0 \qquad (6.28)$$

and

$$\rho \left(\frac{\partial}{\partial t} + \mathbf{v} \cdot \nabla \right) \mathbf{v} = \eta \nabla^2 \mathbf{v} - \nabla \rho. \qquad (6.29)$$

Here $P = P_0 - \rho[(\nabla\phi)^2/(8\pi)][(\partial\varepsilon/\partial\rho)_T]$ with the hydrostatic pressure P_0 and the mass density ρ of the liquid constituting the film.

We should point out that the presence of the surface stresses (Eqs. 6.25 and 6.26) means that the fluid can never stay at rest ($v \neq 0$). Therefore, an absolute threshold of motion does not exist, which is in accordance with the experimental observations. In fact, if the electric field is lower than the threshold value of the VM ($U < U_{cv}$), it generates a slow motion with hydrodynamic flow lines lying in the vertical (y, z) plane (Fig. 6.8).

It is interesting to compare the velocity v_0 of this motion with the velocity v_v of the vortex motion. The order of magnitude of these velocities is given by $v_0 \propto Fh/\eta$ and $v_v \propto 2Fa^2/h\eta$, where F is the average value of the electric surface force. Thus we obtain $v_0/v_v \propto h^2/2a^2 \ll 1$, i.e. the hydrodynamic velocity in the stationary state below U_{cv} is very small (from experiment it is found: $v_0 \approx 10^{-2}$ cm/s and $v_v \approx 10$ cm/s for h/a $\approx 10^{-2}$).

It is not possible to obtain the analytical solution of Eqs. (6.22)–(6.29), even for the stable state below the threshold of the VM. Therefore, one should make some further approximations:

1. The true surface charge density, q, of the regime below U_{cv} is a decreasing function of the y co-ordinate. We will apply the simplest analytical approximation for the $q(y)$ dependence, a linear decreasing function, i.e.

$$\frac{\partial q}{\partial y} = -\mu \frac{U}{a^2}, \qquad (6.30)$$

where μ is a positive dimensionless constant.

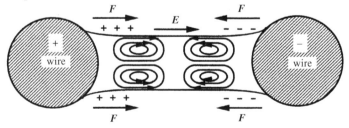

Fig. 6.8 Cross section of a freely suspended nematic film showing the hydrodynamic flow lines for an electric voltage lower than the threshold value, U_{cv}. E is the electric field and F is the surface electric force (reprinted with permission from Reference [5]. Copyright 1983 by the American Institute of Physics)

2. We will use a two-dimensional approach, neglecting any dependence of the various parameters (the electric field, the hydrodynamic velocity, etc.) on the y coordinate. This approximation permits one to neglect the boundary conditions at the conducting wires. Furthermore, we assume, as has been done above for the DM, that the hydrodynamic flow velocity is directed only along the y-axis.

3. The slow hydrodynamic motion, which is present below the threshold of the VM, will be neglected.

The threshold of the VM can be calculated by using the method of linear analysis of the instability already applied for the DM. By substituting Φ, v, q and P, written in the form of Eqs. (6.4)–(6.6), into Eqs. (6.22)–(6.29), and by linearising these equations with respect to $d\Phi$, dv_y, dq and dP, one obtains a set of linear differential equations for $\Phi(z)$, $v(z)$, $q(z)$ and $P(z)$. For convenience we transform these equations into dimensionless ones by defining (a), $(v^2/a(4\pi\sigma)^2)$ and $(1/4\pi\sigma)$ as units of length, mass and time, respectively. These dimensionless equations will have the following form:

$$\frac{\partial^2 \Phi}{\partial z^2} - k_v \Phi = 0, \tag{6.31}$$

$$\frac{\partial^2 \Phi}{\partial z^2} - k_v^2 v = pc_v v, \tag{6.32}$$

$$P = 0. \tag{6.33}$$

Here $k_v = 2\pi/\lambda_v$ is the VM instability wave-vector, c_v is a constant and p is the dimensionless parameter defined as

$$p = \frac{4\pi\rho a^2 \sigma}{\eta}. \tag{6.34}$$

The boundary conditions for the variables $\Phi(z)$, $v(z)$ and q can be obtained by rewriting Eqs. (6.22), (6.23), (6.24) and (6.26) in the new dimensionless units and by linearising these equations with respect to the small fluctuations dq, $d\Phi$ and dv_y. By keeping only linear terms in dq, $d\Phi$ and dv_y, and by using Eq. (6.30), one obtains the following dimensionless boundary conditions on the two free film surfaces ($z = \pm h/2a$):

$$c_v q = -\frac{\nabla\Phi_i \cdot \mathbf{L}}{4\pi} + \mu v, \tag{6.35}$$

$$q = \frac{1}{4\pi}(\varepsilon\nabla\Phi_i - \nabla\Phi_e) \cdot \mathbf{L}, \tag{6.36}$$

$$\Phi_e = \Phi_i, \tag{6.37}$$

$$qE = \frac{1}{R}\frac{\partial v}{\partial L} \tag{6.38}$$

and

$$\Phi_e \to 0 \quad \text{for} \quad z \to \pm \infty. \tag{6.39}$$

Here E is the stationary electric field ($E = 1$ in these dimensionless units), and R is the dimensionless parameter given as

$$R = \frac{U^2}{4\pi\eta\sigma a^2}. \tag{6.40}$$

At the threshold of the VM instability, the fluctuations neither decay nor grow, thus $c_v = 0$. The solutions of Eqs. (6.31) and (6.32) for $c_v = 0$, which satisfy the boundary condition (6.39), are

$$\Phi_e(z) = A \exp(\pm k_v z) \quad (\text{`}+\text{' for } z < -h/2a \text{ and `}-\text{' for } z > h/2a,), \tag{6.41}$$

$$\Phi_i(z) = B \cosh(k_v z) \text{ (for } -h/2a \leq z \leq h/2a), \tag{6.42}$$

$$v(z) = C \cosh(k_v z) \text{ (for } -h/2a \leq z \leq h/2a). \tag{6.43}$$

Here A, B and C are constant.

By substituting these solutions into the boundary conditions (6.35)–(6.38), we will obtain the values of A, B and C and the following expression for the value of R at the threshold:

$$R = R_c = \frac{k_v \left[\sinh\left(\frac{k_v h}{2a}\right) \right]^2}{\mu \left[\cosh\left(\frac{k_v h}{2a}\right) + \varepsilon \sinh\left(\frac{k_v h}{2a}\right) \right] \cosh\left(\frac{k_v h}{2a}\right)}. \tag{6.44}$$

In our experiment $k_v \approx 2\pi$ and $k_v h/a \ll 1$ and thus

$$R = R_c \approx \frac{k_v^3 h^2}{\mu 4 a^2} \propto \frac{2\pi^3 h^2}{\mu a^2}. \tag{6.45}$$

By substituting Eq. (6.45) into Eq. (6.40) we obtain the threshold voltage

$$U_{cv} = 2\pi^2 h \sqrt{\frac{2\eta\sigma}{\mu}}. \tag{6.46}$$

This expression for the threshold voltage does not contain the anisotropies of physical parameters, i.e. it should be valid for both anisotropic and isotropic fluid films. Eq. (6.46) agrees satisfactorily with the experimental data. In particular U_{cv} is proportional to the film thickness, h, and to the square root of the electric conductivity. The proportionality between U_{cv} and $(\eta\sigma)^{1/2}$ can also be obtained by a simple dimensional analysis and thus it holds also if the layer geometry is more complex.

6.3.2 SMECTIC A FILMS

The smectic A phase has very strong anisotropic flow properties. For shears in the plane of the layers, it behaves as a liquid with a viscosity of the order of 1 Poise. In the perpendicular direction it is similar to the plastic crystal. When the orientation is fixed by external boundaries, flow perpendicular to the layers is possible only by the so-called 'permeation' effect (e.g. [17, 18]), in which molecules move from layer to layer. If the layers are not fixed at the boundaries, then perpendicular shears cause the formation of defects in the layer structure and singular lines in the director field. In the smectic A phase, distortions of the director field are always accompanied by distortions of the smectic layers [17, 18].

The EHD instability has been studied in the freely suspended 8CB smectic A film (see Section 4.4.1) [6–8], which smectic layers are spontaneously oriented parallel to the film interfaces. For such a film, its structure fixes the plane of the smectic layers and therefore suppresses any flow normal to the film. At room temperature (23 °C), far from the transition into the nematic phase ($T_{NA} = 33.5$ °C), the 'permeation' effect flow, perpendicular to the layers, is also completely negligible. Flow-induced distortions of the director field are similarly suppressed. The free surfaces of the film have free-slip boundary conditions, so the film can be realistically treated as an isotropic, two-dimensional fluid.

The observed EHD instability is quite similar to the VM in nematic free-standing films, described in the previous paragraph. For example, for both these instabilities, the experimental relation (6.19) is valid, and the proportionality coefficient α has almost the same values. The threshold voltage, U_{cv}, is also linearly proportional to h in both these case (see Eq. 6.46). However, a number of essential differences have also been found. For instance, the flow lines in vortices in smectic films are always parallel or perpendicular to the conducting wires, while inclined vortices are sometimes observed in nematic films (see Fig. 6.5). The vortices in smectic films get smaller when the electric field frequency f becomes greater than several Hz, while in nematic films the vortex dimensions are practically independent of f. All these differences are probably related to the two following circumstances.

(i) Smectic films are more homogeneous in thickness than nematic ones, which makes the former systems more controllable and reproducible;
(ii) The experimentally used film thicknesses are about one order of magnitude smaller for smectics than for nematics (see above in this chapter).

The physical mechanism of the VM hydrodynamic flow in smectic A freely suspended films seems to be most probably similar to that for the VM in nematic films, i.e. the interfacial electric charges should also play an important role in the formation of the vortex patterns in smectic A films.

One of the basic assumptions of the VM theory for nematic films (see previous paragraph) is $h \gg L_D$ (where L_D is the Debye length). This allows one to treat charges, currents and forces as purely surface terms, i.e. to simplify considerably the problem. This simplified treatment, however, is not satisfactory for smectic films, since in this case $h \approx L_D$. A linear stability analysis for the VM in smectic A free-standing films is produced in [19], which is similar in its main features to that described above for the VM in free-standing nematic films, but more developed (accurately taking into account spatial distribution of the electric fields and the interface charges).

6.3.3   SMECTIC C AND C* FILMS

The DM instability in the smectic C and C* free-standing films is predicted theoretically in Reference [20]; however, it has not yet been observed experimentally. The VM instability, similar to that occurring in nematic and smectic A films, is observed in some smectic C and C* freely suspended films [9–11]. It has been found that the vortex instability has a threshold character for smectic C films, while it shows no threshold for smectic C* films. If a smectic C (C*) film, exhibiting vortex instability, is gradually heated into the smectic A phase, the morphology of the vortex flow does not change. This proves that the physical mechanism responsible for the VM instability in smectic A and smectic C (C*) films is the same. Thus the surface electric charges may also play a crucial role in the latter case.

6.4   LINEAR ELECTROPTICAL EFFECT IN SMECTIC C* FREE-STANDING FILMS

A practically applicable electric field-induced phenomenon—linear electrooptical effects—in ferroelectric smectics bounded by solid substrates has already been discussed briefly in Chapter 1. This effect in free-standing smectic C* films exhibits some peculiarities compared to that in the solid plate-bounded samples (e.g., [21–25]). These differences are determined by the non-identical boundary conditions for freely suspended films and solid-confined samples: the degenerated boundary conditions in the first case and the fixed director orientation at the surfaces in the second case; and by the geometrical dimensionality—the possibility of existence of the two-dimensional free-standing films.

Thus polarisation-optical and ellipsometric studies show that the linear electro-optical-effect threshold field (E_{th}) (see Eq. 1.24) is negligible and the response time (t_r) is about two orders of magnitude shorter in free-standing smectic C* films of (R)-4'-(1-butoxycarbonyl-1-ethoxy) phenyl 4-[4-(n-octyloxy)phenyl] benzoate (1BC1EPOPB) (with thicknesses of the order of several tens of molecular layers), compared to the 1 μm-thick samples of the same material bounded by solid plates [21]. Note, finally, that the spontaneous polarisation P_{sp} in smectic C* free-standing films is found to increase with the diminishing film thickness, which is due to the ordering effect of the film interfaces [24, 25].

REFERENCES

1. Meyerhofer D., Sussman A., Williams R., 1972, *J. Appl. Phys.*, **43**, 3685.
2. Faetti S., Fronzoni L., Rolla P.A., Stoppini G., 1976, *Lett. Nuovo Cimento*, **17**, 475.
3. Faetti S., Fronzoni L., Rolla P.A., 1979, *J. Phys. (France) Colloq.*, **40**, C3–497.
4. Faetti S., Fronzoni L., Rolla P.A., 1983, *J. Chem. Phys.*, **79**, 1427.
5. Faetti S., Fronzoni L., Rolla P.A., 1983, *J. Chem. Phys.*, **79**, 5054.
6. Morris S.W., de Bruyn J.R., May A.D., 1990, *Phys. Rev. Lett.*, **65**, 2378.
7. Morris S.W., de Bruyn J.R., May A.D., 1991, *Phys. Rev. A*, **44**, 12.
8. Morris S.W., de Bruyn J.R., May A.D., 1991, *J. Stat. Phys.*, **64**, 1025.
9. Becker A., Ried S., Stannarius R., Stegemeyer H., 1997, *Europhys. Lett.*, **39**, 257.
10. Langer Ch., Stannarius R., Becker A., Stegemeyer H., 1998, *Proceedings of SPIE*, **3318**, 154.
11. Langer Ch., Stannarius R., 1998, *Phys. Rev. E* (in press).
12. Advsec D., Luntz M., 1937, *C. R. Acad. Sci.*, **204**, 797.
13. Blinov L.M., 1983, *Electro-Optical and Magneto-Optical Properties of Liquid Crystals*, Wiley, Chichester, UK (Russian version: 1978, Nauka, Moscow).
14. Blinov L.M., Chigrinov V.G., 1994, *Electrooptic Effects in Liquid Crystal Materials*, Springer, New York.
15. Chandrassekar S., 1961, *Hydrodynamic and Hydromagnetic Stability*, Clarendon Press, Oxford, UK.
16. Marangoni G., 1872, *Nuovo Cimento*, **2**, 239.
17. de Gennes P.G., 1974, *The Physics of Liquid Crystals, Clarendon Press, Oxford, UK.*
18. de Gennes P.G., Prost J., 1993, *The Physics of Liquid Crystals*, 2nd edn., Clarendon Press, Oxford, UK.
19. Daya Z.A., Morris S.W., de Bruyn J.R., 1997, *Phys. Rev. E*, **55**, 2682.
20. Ried S., Pleiner H., Zimmermann W., Brand H.R., 1995, *Phys. Rev. E*, **53**, 6.
21. Uto S., Ohtsuki H., Terayama M., Ozaki M., Yoshino K., 1996, *Jpn. J. Appl. Phys.*, **35**, L158.
22. Becker A., Stegemeyer H., 1997, *Liq. Cryst.*, **23**, 463.
23. Becker A., Stegemeyer H., 1997, *Ber. Bunsenges. Phys. Chem.*, **101**, 1957.
24. Hoffmann E., Stegemeyer H., 1996, *Ber. Bunsenges. Phys. Chem.*, **100**, 1250.
25. Hoffmann E., Stegemeyer H., 1996, *Ferroelectrics*, **179**, 1.

7

Structure and Defects of Stable Thin Films

The main principles of molecular organisation of some FSLC films have already been briefly described in Chapter 2. Here, on the basis of the structural experimental data, we will analyse in detail the molecular structure and topological defects of some thermodynamically stable freely suspended films. These are (as we already know, see Chapter 3) the films that do not considerably change their thickness, molecular organisation and physical properties in a certain time interval (sufficient for experimental determination of the film structure). Such 'stable' films, considered below, are thermotropic smectic and black soap films, as well as biological membranes (see also Chapters 3, 4 and 8).

7.1 STRUCTURE

7.1.1 THERMOTROPIC SMECTIC FILMS

One may obtain thermodynamically stable thermotropic smectic films in a large interval of thicknesses: from many thousands of molecular layers to one bilayer. In addition, as we know (see Chapter 5), such films exhibit rich phase metamorphism. All this makes smectic films the promising systems for structural studies: light scattering, X-ray reflection and scattering, electron diffraction, neutron reflection, etc. (see Chapter 3).

The smectic A free-standing films possess the most simple molecular organisation. As an example, consider the molecular structure of free-standing films of $4'$-n-alkyl-4-cyanobiphenyls: 8CB and 12CB (see Sections 4.4.1 and 5.3.3), exhibiting smectic A phase at room temperature [1]. Fig. 7.1 represents the schematic electron density profiles $\sigma(z)$ for such films, in the direction perpendicular to the molecular layers, obtained by means of X-ray reflection. Two different situations are possible:

1. the electron density modulation period δ is commensurate with the film thickness h. In this case the film contains an integer number of undisturbed smectic monolayers, Fig. 7.1(a).

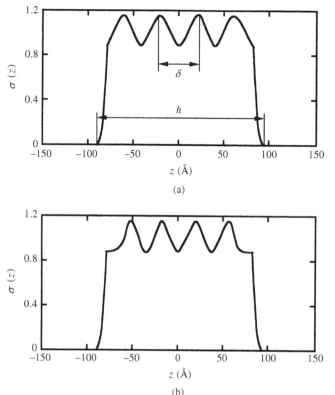

Fig. 7.1 Schematic co-ordinate distributions of the electron densities for freely suspended smectic A films, calculated from X-ray reflectivity measurements data (reproduced by permission of EDP Sciences from Reference [1]): (a) commensurate four-layer film of 12CB at 55 °C; (b) incommensurate four-layer film of 8CB at 23 °C

2. δ is incommensurate with h. The most probable structure of the film in this case is an integer number of undisturbed smectic monolayers in the bulk, with some additional material at both film interfaces, Fig. 7.1(b).

7.1.2 BLACK SOAP FILMS

Intermolecular forces in soap films involve the basic physical interactions existing in more complex structures, for example, biological membranes. This explains the structural similarity of both these media and makes the study of molecular organisation of soap films a useful tool to obtain information on biological systems.

As we know, there exist two types of thermodynamically stable thin soap films: Newton black films (NBF) and common black films (CBF) (see Chapter 4). For the NBF, strongly repulsive short-range forces associated with the local structure of water (hydration, steric, undulation and 'protrusion' [2]) are expected to play a role in the stability. For the CBF, the equilibrium thickness is known to be determined by a balance between the van der Waals attraction and the double-layer repulsion forces [3]. The NBF are formed when the impurity salt concentration, c, in a soap solution is quite high. In this case, the electric field from the salt ions screens the double-layer attraction forces, so the film can reach lower equilibrium thicknesses. The CBF are normally formed for low concentrations of salt ($c < 0.1$ mol/l).

A first attempt to study the structure of a soap black film (prepared from a decyltrimethylammonium decyl sulphate solution) in the direction perpendicular to the film interfaces was made by means of X-ray scattering, using a standard X-ray diffractometer (in Reference [4]). However, these measurements were not very sensitive and led to a very overestimated value of the total thickness of the film (ranging from 50 to 70 Å), and thus an incorrect interpretation of the water layer thickness (20 Å).

The only (to our knowledge) reported neutron reflectivity experiment [5] is related only to relatively thick soap films (n-decyl trimethylammonium bromide mixed with decanoic acid and D_2O), since the authors were unable to maintain the CBF long enough to produce reliable reflectivity profiles (see also Chapter 3).

The most detailed and comprehensive study of the black soap film structures was made by means of X-ray reflectivity in Reference [6] (see also Chapter 3). A complete molecular structure was determined in the direction perpendicular to the film interfaces (i.e. the exact amount of water between the amphiphilic layers, the interfacial roughness, the different densities, etc.) of NBF and CBF, made from sodium dodecyl sulphate (SDS) solutions in the presence of a salt (NaCl).

a   Newton Black Films

A schematic representation of the structure of a NBF with the SDS concentration equal to 1 g/l (higher than CMC) and with the NaCl concentration equal to 0.4 mol/l, is shown in Fig. 7.2(a). This NBF is relatively stable (its mean lifetime is more than 45 min). The film consists of two tilt and, possibly, stretched tails of amphiphilic monolayers (aliphatic medium), divided by two layers of polar surfactant heads and a thin aqueous core. The total thickness of this NBF is found to be 32.9 ± 0.5 Å. This result is important, since it means that the NBF is so thin that it cannot contain any water in a liquid state, as opposed to the previous X-ray [4] and optical [7] measurements. The aqueous core thickness obtained from the experimental data (3.7 ± 0.5 Å) corresponds to a hydration layer of the polar heads of the surfactant molecules, or more exactly to sodium ions with hydration shells in order to preserve the electroneutrality. The determined values of roughness, densities, and chain length of the aliphatic medium of the film are found to be, respectively,

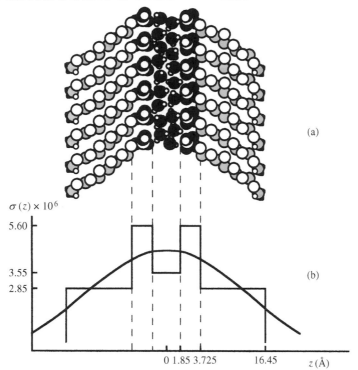

Fig. 7.2 (a) Schematic view of the structure of a Newton black film. (b) Profile of the electron density along the normal to the film interfaces (reproduced with permission from Reference [6]. Copyright 1991 the American Physical Society)

$v = 2.7 \pm 0.1\,\text{Å}$, $\delta_{ch} = (2.85 \pm 0.1) \times 10^{-6}$ and $l_c = 10.85 \pm 0.5\,\text{Å}$. The calculated area per surfactant molecule is found to be $33\,\text{Å}^2$ [6].

The corresponding electron-density profile $\sigma(z)$ is represented in Fig. 7.2(b).

b Common black films

The CBF of SDS are obtained in Reference [6] for low (0.1 mol/l) NaCl concentrations. For such a low salt content, the film is clearly unstable, and its mean lifetime becomes shorter (≈ 1–5 min). Moreover, as the film now contains water, the corresponding 'noise' in the X-ray reflected signal is increased. Nevertheless, it has been possible to estimate the total thickness of such a film: $54.4 \pm 0.5\,\text{Å}$. This value is significantly smaller than the CBF thickness reported in Ref. [7] ($\approx 90\,\text{Å}$). The thickness of the aqueous core is found to be $27 \pm 0.5\,\text{Å}$. This means that it does not represent the hydration layer of the polar heads, as in the case of the NBF, but clearly a liquid layer.

For the CBF, practically the same as for the NBF are the values of roughness, densities, and chain length of the aliphatic medium (respectively, $v = 2.75 \pm 0.15$ Å, $\delta_{ch} = (2.94 \pm 0.1) \times 10^{-6}$ and $l_c = 10.3 \pm 0.5$ Å) and, consequently, almost the same as for the NBF is the area per surfactant molecule (≈ 34 Å^2) [6].

This result clearly demonstrates that the molecular organisation is practically identical for both NBF and CBF, the difference in their structures consisting only of a greater value of the aqueous-core thickness for CBF.

Note, finally, that structures of NBF and CBF with non-tilted surfactant mono-layers are, in principle, also possible.

c   Transition between Newton and common black films

The question of the nature of the transition between NBF and CBF has not yet been completely elucidated. Nevertheless, it has been observed previously [8] that it is possible to have the two black films coexisting within the same film. However, the data of Reference [6] show that, at a given surfactant solution concentration and temperature, the film drawn under the same conditions will form either the NBF or the CBF. In any case, as is clear from the just-described data on the structure of NBF and CBF, the NBF–CBF transition should consist only of a strong and abrupt increase of the aqueous-core thickness, whereas the aliphatic medium structure should remain unchanged.

The interested reader can find more information on experimental studies on black soap films (for instance, on the issue of the black film stability; on the behaviour of the edge angle between the black film and the surrounding meniscus in the dependence upon the salt concentration, as a criterion of the CBF–NBF transition, etc.) in Reference [9].

7.1.3   BIOMEMBRANES

The just-acquired notions on the molecular organisation of black soap films will help us now to analyse the structure of more complex objects—biological membranes. Indeed, an isolated biomembrane can be considered as a freely suspended liquid film, containing bilayers of surfactant molecules as one of the main structural components (e.g. [3, 10, 11]).

Membranes are the most common cellular structures for both animals and plants. They are involved in all kinds of the life activity: food entrapment and transport, immunological recognition, nerve conduction, biosynthesis, etc.

The molecular organisation of biological membranes is very complex and varied. Each membrane commonly contains 50 or more different proteins and a host medium of phospholipids or glycolipids with various head groups, numbers of chains, chain lengths, as well as steroids (e.g. cholesterol) and other amphiphilic molecules. This is why we will restrict ourselves here only to a brief and schematic description of the membrane organisation.

Consider firstly the structure of membrane lipids. Most biological membrane lipids are double-chained phospholipids or glycolipids. Each of these hydrocarbon chains contains from 16 to 18 carbon atoms. One chain is saturated, being formed from saturated chemical bonds, while the other one is unsaturated, and contains 1–3 double chemical bonds.

Nature has chosen such lipids to be the principal constructive blocks of biomembranes not accidentally, but to assure the following properties of membranes [3].

(i) Biological lipids can easily self-assemble into thin bilayers (membranes), which can compartmentalise different regions within a cell and protect the inside of the cell from the outside.

(ii) Due to the extremely low lipid CMC, the membranes remain intact even when their exterior bathing medium is strongly depleted of lipids.

(iii) Because of the unsaturation of one of the lipid chains, the membranes are always in the fluid state at physiological temperatures.

The most common lipids in animal cells are the two phospholipids: phosphatidylcholine (PC) and phosphatidylethanolamine (PE), while in plant cells they are the glycolipids: digalactosyl diglyceride (DGDG) and monogalactosyl diglyceride (MGDG). In each case the first lipid is packed in a truncated cone, while the second is packed as an inverted truncated cone. Thus depending on the ratio of these lipid pairs in a bilayer, they can be organised into planar bilayers [Fig. 7.3(a)], or into bilayers of varying curvatures and flexibilities.

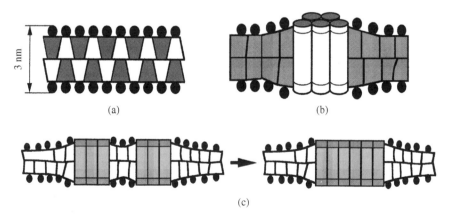

Fig. 7.3 (a) Mixture of two different lipids packed together in a planar membrane. One of these lipids (e.g. lecithin or DGDG) is packed as a truncated cone, the other (e.g. PE, MGDG, cholesterol) is packed as an inverted truncated cone. (b, c) Packing constraints induced in the hydrocarbon chain regions of lipids around a protein molecule, which may be relaxed when proteins tilt (not shown) or aggregate, as illustrated in (c) (adapted from Reference [3])

In biomembranes, different proteins are usually incorporated into lipid bilayers [Fig. 7.3(b)]. Proteins are long-chained polypeptides—polymers consisting of a long string of amino acid residues. Compared to lipids, membrane proteins are structurally rigid. The protein chains are folded into a number of cylindrical segments, each exposing a hydrophobic surface. These segments are linked together by hydrophilic residues [smaller cylinders in Fig. 7.3(b)]. The hydrophobic helices span the lipid bilayer into which they are incorporated, while the hydrophilic residues occupy the bilayer–water interfacial regions on both sides of membranes.

When proteins are incorporated into a lipid bilayer, they usually disturb the lipid organisation in their neighbourhood. Such distorted lipids are called the 'boundary lipids' [see Fig. 7.3(b)]. Their perturbation is due to the different geometrical sizes of proteins and lipids and to the specific character of interaction forces acting between them (see, for example, Reference [3]). The distorted lipid chain region near the protein molecules is often mechanically strained. The constraints on the lipid packing may be relieved if proteins tilt or aggregate, as shown in Fig. 7.3(c).

Note that the just-considered organisation of a biological membrane is to a high degree idealised. In reality, a biomembrane is a dynamic structure: both the lipids and the proteins move about rapidly in the plane of the membrane. However, heterogeneous domains and local clustering of lipids and proteins (analogous to those shown in Fig. 7.3) also occur, which is important for the normal chemical functioning of the membrane and its components (see, for example, Reference [3]).

The interested reader can find a wide range of information on biomembranes structure and function, for instance, in the books in Reference [3, 12–14]. Some issues concerning membranes functional activity will also be briefly discussed in the following chapter.

7.2 DEFECTS

In principle, FSLC films (and especially the thick ones) should exhibit the whole variety of topological defects observed in the bulk mesophase samples (see also Chapters 1, 5 and 6). For example, the 180°-walls and disclinations already described in Chapter 1 may occur in thick and even in thin (two-dimensional) free-standing smectic C* films (e.g. [15, 16]). These defects disappear under the action of an external electric field. The ordering effect of the free film interfaces (which are more pronounced in thin films) may also suppress topological peculiarities, making the film texture homogeneous. For example, freely suspended films of thermotropic nematic 5CB exhibit homogeneous homeotropic orientation even for quite great thicknesses, of the order of $100\,\mu m$ (see Section 5.1.1).

In addition, a number of specific, purely interfacial (or two-dimensional) defects may appear in thin (two-dimensional) FSLC films: boojums, stripes, targets, spirals, etc.

The presence of defects may influence certain thermodynamic properties of FSLC films, for instance the phase transitions temperatures.

We will consider in more detail some of the topological peculiarities in FSLC films.

7.2.1 TWO-DIMENSIONAL DEFECTS

Defects are especially variable and unusual in chiral smectic films, due to the complexity of the boundary conditions in this case. An example is the defect textures observed experimentally in a free-standing film of *R*-hexyloxybenzylidene *p'*-amino-2-chloropropyl cinnamate (HOBACPC), a chiral liquid crystal which forms ferro-electric smectic C* and smectic I phases [17]. In the smectic C* phase, the usual schlieren texture already mentioned in Chapter 5 is observed in the crossed nicols of a microscope. This texture is common for both bulk mesophase samples and FSLC films, with discontinuously degenerated (planar or tilted) boundary conditions in the absence of the direction of the preferred or 'easy' orientation (see, for example, References [18, 19]).

On cooling this film to a temperature of about 55 °C, one obtains purely two-dimensional topological defects. Firstly, small, roughly circular domains of smectic I nucleate. These smectic I droplets drift away from their nucleation sites and transform into smectic C* [Fig. 7.4(a)]. The droplets are characterised by from one to three schlieren lines which appear to originate at a point at the edge of the droplet. The schlieren lines lie at 45 ° to one another. The mutual orientation of schlieren lines and polarisers indicates that the liquid crystalline optical axis is either parallel or perpendicular to the droplets boundaries. Under further cooling, the smectic I droplets grow and merge with one another, filling the whole area of the film and forming a striped pattern [Fig. 7.4(b)].

The schlieren lines in this latter texture indicate that the optical axis is parallel to the edge of the stripe along the edge. Moving across the stripe the optical axis rotates through 180 °, until it is parallel to the opposite edge of the stripe. At this point there is a defect line at which the optical axis reverses itself and a new stripe begins. The direction of rotation across the stripe is always the same. The striped pattern belongs to periodic modulated textures already mentioned in Chapter 5. Note that similar striped pattern textures were also observed in non-chiral smectic C free-standing films [20].

The above-described defect textures are characteristic only of chiral films. If the same experiment is produced using a racemic non-chiral mixture of HOBACPC (equal numbers of right- and left-handed molecules), the texture will be uniform, with no converging schlieren lines and stripes.

The presence of the striped and schlieren droplet patterns for a chiral film and the absence of these patterns for a non-chiral film can be explained from the point of view of the different symmetries of these systems. De Gennes [18, 19] has suggested that a tilted smectic in three dimensions might exhibit a finite density of defects due

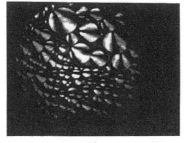

(a)

(b)

Fig. 7.4 Defects in the HOBACPC free-standing film in the vicinity of the smectic C*–smectic I phase transition, viewed in a polarising microscope (reproduced with permission from Reference [17]. Copyright 1986 by the American Physical Society): (a) smectic I droplets with schlieren lines; (b) smectic I with the striped texture

to the chirality of its molecules. The tilted HOBACPC film is a two-dimensional realisation of this prediction.

A tilted smectic film, as we know, may be characterised by the two-dimensional director, c (see Eq. 5.29). Then the free energy, Φ, of such a film may be represented in a general form as [17]

$$\Phi[c] = \int \int \mathrm{d}x \; \mathrm{d}y \; F[c(r)] + \Phi_{\mathrm{defect}}, \qquad (7.1)$$

where F is the free energy density (a scalar function of c and its gradients), r is the radius-vector in the film plane, and Φ_{defect} is the free energy of defect cores.

There are two contributions to F: bulk terms involving integrals of gradients of c over the area of the system, and surface terms involving line integrals around the boundary of the system. In a system of size R the gradients of the order parameter scale as $1/R$. Since the bulk free-energy density is quadratic in the gradients, the total bulk free energy (integrated over the sample area) will scale independently of R. On the other hand, the surface free energy per unit length is independent of the system size, so the total free energy will scale as R. Hence, for large systems, where R is much greater than the molecular length, the surface energy will dominate over the bulk one. Strong surface energies imply a strong preferred angle between c and the boundary. These strong pinning boundary conditions are crucial in determining the patterns that the film will display.

The assumption that the smectic I droplets are perfectly circular (due to the surface tension of the smectic C*–smectic I interface) and that c lies parallel to the

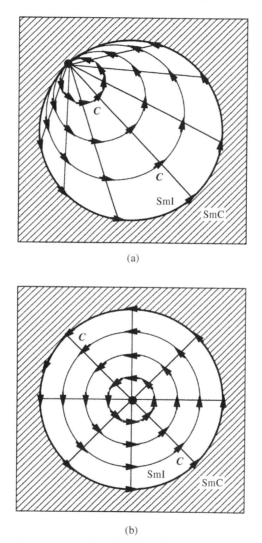

(a)

(b)

Fig. 7.5 Schematic views of the director field inside a smectic I droplet with schlieren lines [see Fig. 7.4(a)], representing the two-dimensional analogues of boojums (reproduced with permission from Reference [17]. Copyright 1986 by the American Physical society): (a) a point defect at the droplet boundary; (b) a point defect in the droplet centre

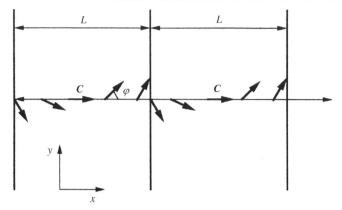

Fig. 7.6 Schematic director distribution inside a smectic I free-standing film with a striped texture (see Fig. 7.4b) (reproduced with permission from Reference [17]. Copyright 1986 by the American Physical Society)

interface, lead to the minimum-energy droplet textures shown in Fig. 7.5. These droplets give the straight schlieren lines situated at $45°$ to one another. The first droplet [Fig. 7.5(a)] contains the point defect placed at its boundary. This defect can be expelled from the droplet at a small cost in surface energy, but a great saving in bulk gradient energy. The second droplet [Fig. 7.5(b)] contains a point defect at its centre.

Both these droplets are two-dimensional analogues of the famous boojum in superfluid ^{3}He-A [21].

Similar defects were observed in spherical, liquid crystalline drops suspended in some amorphous fluids (e.g. [22, 23]).

The striped texture is determined by the competition between the bulk and the surface free energies. The bulk energy is minimised by a uniform order parameter, but the boundary terms prefer c to be pointed in opposite directions on opposite sides of the stripe. Both conditions can be satisfied arbitrarily well in arbitrarily large stripes. However, if the energetic cost of a defect line is not too great, the system prefers to gain bulk gradient energy in order to lose surface energy. Thus the stripes will have a finite width, L. A schematic director distribution for such a striped pattern, corresponding to the experimental observations and giving a minimum to the free energy is represented in Fig. 7.6.

The just-described boojum droplets and striped patterns are quite common for free-standing chiral smectic films (e.g. [24–27]). However, some other two-dimensional defect-containing textures may also occur in these films (e.g. [24–30]). Fig. 7.7 shows an example of the modulated hexagonal texture composed of intersecting walls (W) and disclinations, lying at the centre of the hexagonal cell (L_1) and at its corners (L_2), which is predicted theoretically in References [24, 25]. Similar, but even more complex modulated textures are observed in crystalline smectic B, 7O.7, free-standing films already mentioned in Chapter 5 [30]. A number of peculiar defect structures occur in the smectic C* free-standing films of 4-(3-methyl-2-chloropentanoyloxy)-4'-heptyl-oxybiphenyl (C7) [29]. These are, for instance, unstable point defects and the so-called 'chessboard' texture (Fig. 7.8).

Another beautiful example of the two-dimensional defect pattern, typical for all circular-form free-standing smectic films, is the so-called arch texture (Fig. 7.9(a) [31]. This is a stable (non-stratifying) stepwise structure, consisting of several concentric circular portions of the film divided by the edge dislocations. Thicknesses of the neighbouring portions differ by an integer number of smectic layers. A simplified schematic structure of a portion of this defect pattern [Fig. 7.9(b)] is similar to that of the stratifying swollen lamellar phase film [see Fig. 4.10(d)].

An interesting issue is the thermal diffusion (Brownian) motion of defects in FSLC films. Indeed, topological peculiarities in such films slowly diffuse with time over the film interfaces, as is the case with small solid particles suspended on a soap (micellar) film surface [32], or a fluorescent substance diluted in a smectic A (8CB) film [33]. This Brownian motion was observed experimentally for wedge disclination [shown in Fig. 1.5(c)] in a three-layer-thick smectic C free-standing film of 4-[(S)-(4-methylhexyl)oxy]-phenyl 4-(decyloxy)benzoate (10.O7) [34]. The corresponding diffusion constant, D_s, was estimated to be of the same order of magnitude ($\approx 10^{-8}$ cm^2/s) as for particles and fluorescent impurities in soap and smectic A films [32–34].

Note that the Brownian motion of defects cannot be observed in thick (three-dimensional) free-standing films. Indeed, the disclination lines are effectively under a tension that has the value of a typical nematic Frank constant, $K \approx k_B T/a$ (where k_B is the Boltzmann constant, T is the temperature and a is the molecular length). A simple calculation shows that the local mean-square fluctuation in the position of a line of length l in three dimensions will be $\langle u^2 \rangle = la$, which is not resolvable until l is

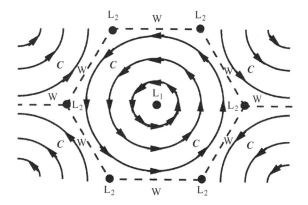

Fig. 7.7 Schematic representation of the director field for the hexagonal texture of a chiral smectic free-standing film. Disclination lines, L_1 and L_2, and walls, W, are perpendicular to the figure plane (adapted with permission from References [24, 25]. Copyright 1988, 1989 by the American Physical Society)

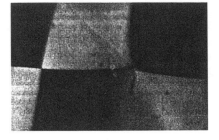

(a)

(b)

Fig. 7.8 (a) The 'chessboard' texture in a smectic C* free-standing film of the substance C7 (viewed in a polarising microscope with slightly uncrossed polarizers). (b) The corresponding director distribution. Disclination line, L, and walls, W, are perpendicular to the figure plane (reproduced by permission of Taylor & Francis from Reference [29])

~ 1 cm or greater. By using a thin free-standing film, one effectively takes a slice of the nematic of thickness h, giving $\langle u^2 \rangle = la(l/h)$, which enhances $\langle u^2 \rangle$ by the large ratio l/h, so that the fluctuations are readily observable. Moreover the defects Brownian motions cannot be detected in liquid crystalline samples bounded by solid substrates, since, in this case, the surface disclinations, are normally trapped by the solid boundary defects (e.g. [23, 34]).

All the above-described two-dimensional defect patterns are static, i.e. they are not accompanied by a pronounced fluid flow in their vicinity. However, the dynamic defects, which are initiated by a liquid flow, can also appear in free-standing films. It happens, for instance, in the case of the EHD instabilities in FSLC films already described in Chapter 6, or in the case of vortex motion initiated in smectic C and C* free-standing films by thermally induced gradients of surface tension [35].

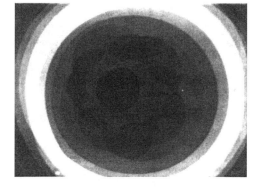

(a)

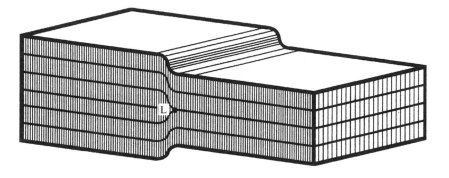

(b)

Fig. 7.9 (a) The arch texture in a smectic A freely suspended film of the substance S2 (observed in a reflecting microscope). (b) A simplified schematic structure of this texture portion. L is the edge dislocation line (reproduced from Reference [31] by permission of Elsevier Science)

Another interesting example of the dynamic two-dimensional topological singularities are two coexisting defect patterns (the so-called 'spirals' and 'targets') observed in thick ($h \approx$ several µm) free-standing circular-form smectic C films of 4'-hexyloxyphenyl 4-decyloxybenzoate (10E6). These patterns appear under the action of a shear flow (which is induced by the rotation of a thin glass needle placed in the film centre) [36, 37], or of a rotating electric field (modulated by two mutually orthogonal pairs of the in-plane electrodes) [38] (Fig. 7.10). Note that similar disclination patterns are characteristic of many other phenomena: Rayleigh–Bénard thermal convection in gases (e.g. [39]), autocatalytic chemical reactions (e.g. [40]), etc.

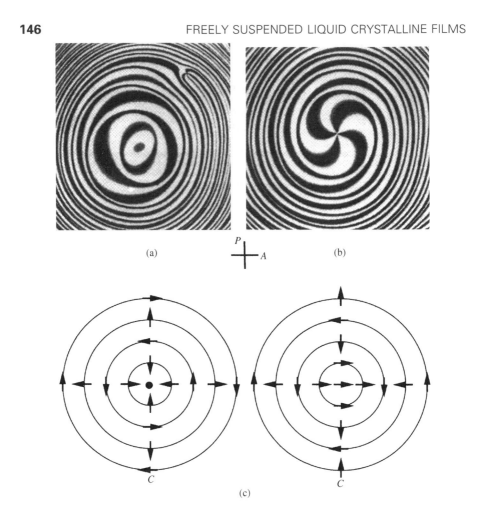

Fig. 7.10 (a, b) Defect patterns obtained in a thick circular-form free-standing smectic C film of the 10E6 substance (viewed in a polarising microscope; the directions of the analyser and the polariser are shown as *A* and *P*, respectively): (a) target; (b) spiral; (c) examples of in-plane director *c* distributions giving the contrast between crossed polarisers which is characteristic of the target and spiral patterns (reproduced with permission from Reference [38]. Copyright 1995 by the American Physical Society)

7.2.2   DEFECTS AND PHASE TRANSITIONS

One should expect an influence of defects on the phase transition temperature, for example, in the case of the smectic A–nematic phase transfer in a free-standing film, containing edge dislocation loops in the smectic A phase. Such loops were recently obtained experimentally by local heating of the smectic A film interfaces [41]. As was shown in Reference [42], the smectic A–nematic phase transition could be initiated by the unbinding of the dislocation loops. The mechanism would be similar to the one proposed by Kosterlitz and Thouless for phase transfers in two dimensions [43, 44] (see also Chapter 5).

The bulk free energy associated with distortions of the lamellar or smectic phase is given by [18, 19] (see Eq. (1.16))

$$\Phi_b = \frac{1}{2} \int d\mathbf{r}_\perp \int dz \left\{ B\left[\frac{\partial u(\mathbf{r})}{\partial z}\right]^2 + K[\Delta_\perp u(\mathbf{r})]^2 \right\}, \tag{7.2}$$

where $u(\mathbf{r})$ is the vertical displacement of the layers from their equilibrium position at $\mathbf{r} = (\mathbf{r}_\perp, z) = (x, y, z)$; $\Delta_\perp$ is the two-dimensional Laplacian with respect to x, y; and B and K are the smectic elastic constants associated with layers compression and splay, respectively.

The surface part of the free energy can be written as [42, 45]

$$F_s = \frac{1}{2} \int d\mathbf{r}_\perp [\gamma |\nabla u_s|^2 + K_s (\Delta_\perp u_s)^2]. \tag{7.3}$$

Here $u_s = u(\mathbf{r}_\perp, z = 0)$, γ is the surface tension, and K_s is the surface splay elastic constant.

In the simplest approximation, the free energy of the dislocation loop in the bulk is given by [42]

$$F_{\text{loop}} = F_c L - k_b T \frac{L}{d} \ln p. \tag{7.4}$$

Here $F_c = \sqrt{(KB)} l^2 / 2r_c + E_c$ is the self-energy of the dislocation per unit length, l is the layer spacing, r_c is the radius of the dislocation core, E_c is the core energy, L is the total length of the loop, p is a number larger than 1 (for the problem considered on the lattice this number would be close to the co-ordination number of the lattice).

Below the bulk phase transition temperature, $T_{\text{NA}}(\infty) = F_c l / (k_B B \ln p)$, F_{loop} changes sign and the growth of the loop is favourable.

Minimising the total free energy $F = F_b + F_s$ (see Eqs. 7.2 and 7.3) with respect to u, one can calculate the dislocation distortion energy per unit length as a function of the film thickness, $F_c(h)$, and substitute its value into Eq. (7.4). For a dislocation loop of circular form with radius R in its centre ($R \gg l$) it gives the phase transition temperature $T_{\text{NA}}(h)$ in the film will be related to $T_{\text{NA}}(\infty)$ as [45]

$$\frac{T_{\text{NA}}(h) - T_{\text{NA}}(\infty)}{T_{\text{NA}}(\infty)} = \frac{2r_c}{\sqrt{\lambda h}}. \tag{7.5}$$

Here $\lambda = \sqrt{K/B}$, i.e. $T_{\text{NA}}(h)$ grows with the diminished h and increased r_c. For a three-layer film, for example, $h = 3l$, $\lambda \approx l = 30 \, \text{Å}$ and r_c equals the width of one molecule (the choice corresponding to zero dislocation core energy), $r_c \approx 3 \, \text{Å}$. Then, from formula (7.5), we obtain that the transition temperature can be shifted by tens of degrees upwards in comparison to the bulk. This means that it should be relatively easy to overheat a smectic film, a fact also observed experimentally [46].

REFERENCES

1. Gierlotka S., Lambooy P., de Jeu W.H., 1990, *Europhys. Lett.*, **12**, 341.
2. Israelachvili J.N., Wernnerström H., 1990, *Langmuir*, **6**, 873.
3. Israelachvili J.N., 1992, *Intermolecular and Surface Forces*, 2nd edn., Academic Press, London.
4. Clunie J.S., Corkill J.M., Goodman J.F., 1966, *Discuss. Faraday Soc.*, **42**, 34.
5. Highfield R.R., Humes R.P., Thomas R.K., Cummins P.G. *et al.*, 1984, *J. Colloid. Interface Sci.*, **97**, 367.
6. Bélorgey O., Benattar J.J., 1991, *Phys. Rev. Lett.*, **66**, 313.
7. Jones M.N., Mysels K.J., Sholten P.C., 1966, *Trans. Faraday Soc.*, **62**, 1336.
8. Duyiss E.M., Overbeeck J.Th.G., 1962, *Proc. K. Ned. Akad. Wet., Ser. B*, **65**, 26.
9. Kruglyakov P.M., Ekserova D.R., 1990, *Foam and Foam Films*, Moscow, Khimia (in Russian).
10. Kruglyakov P.M., Rovin Yu.G., 1978, *Physical Chemistry of Black Hydrocarbon Films. Biomolecular Lipid Membranes*, Nauka, Moscow.
11. Dukhin S.S., Ruleuv N.N., Dimitrov D.S., 1986, *Coagulation and Dynamics of Thin Films*, Kiev, Naukova Dumka (in Russian).
12. Harrison R., Lunt G.G., 1980, *Biological Membranes*, 2nd edn., Halsted-Wiley, New York.
13. Evans E.A., Skalak R., 1980, *Mechanics and Thermodynamics of Biomembranes*, CRC Press, Boca Raton, FL.
14. Robertson R.N., *The Lively Membranes*, 1983, Cambridge University Press, London.
15. Pindak R., Young C.Y., Meyer R.B., Clark N.A., 1980, *Phys. Rev. Lett.*, **45**, 1193.
16. Kim B., Lee S.J., Lee J.-R., 1996, *Phys. Rev. E*, **53**, 6061.
17. Langer S.A., Sethna J.P., 1986, *Phys. Rev. A*, **34**, 5035.
18. de Gennes P.G., 1974, *The Physics of Liquid Crystals*, Clarendon Press, Oxford, UK.
19. de Gennes P.G., Prost J., 1993, *The Physics of Liquid Crystals*, 2nd edn., Clarendon Press, Oxford, UK.
20. Pang J., Clark N.A., 1994, *Phys. Rev. Lett.*, **73**, 2332.
21. Mermin N.D., 1977, in *Quantum Fluids and Solids* (eds. S.B. Trickey, E. Adams, J. Duffy), Plenum, New York.
22. Volovik G.E., Lavrentovich O.D., 1983, *Zh. Eksp. Teor. Fiz.*, **85**, 1897.
23. Sonin A.A., 1995, *The Surface Physics of Liquid Crystals*, OPA-Gordon and Breach, Amsterdam.
24. Hinshaw G.A., Jr., Petschek R.G., Pelcovits R.A., 1988, *Phys. Rev. Lett.*, **60**, 1864.
25. Hinshaw G.A., Jr., Petschek R.G., 1989, *Phys. Rev. A*, **39**, 5914.
26. Chen C.-M., MacKintosh F.C., 1995, *Europhys. Lett.*, **30**, 215.
27. Ohyama T., Jacobs A.E., Mukamel D., 1996, *Phys. Rev. E*, **53**, 2595.
28. Demikhov E.I., 1995, *Pis'ma Zh. Eksp. Teor. Fiz.*, **61**, 951.
29. Demikhov E.I., Stegemeyer H., 1995, *Liq. Cryst.*, **18**, 37.
30. Sirota E., Pershan P.S., Deutsch M., 1987, *Phys. Rev. A*, **36**, 2902.
31. Pieranski P., Beliard L., Tournellec J.-Ph., Leoncini X. *et al.*, 1993, *Physica A*, **194**, 364.
32. Cheung C., Hwang Y.H., Wu X.-I., Choi H.J., 1996, *Phys. Rev. Lett.*, **76**, 2531.
33. Bechhoefer J., Géminard J.-C., Bocquet L., Oswald P., *Phys. Rev. Lett.*, **79**, 4922.
34. Muzny Ch.D., Clark N.A., 1992, *Phys. Rev. Lett.*, **68**, 804.
35. Godfrey U.I., van Winkle D.H., 1996, *Phys. Rev. E*, **54**, 3752.
36. Cladis P.E., Couder Y., Brand H.R., 1985, *Phys. Rev. Lett.*, **55**, 2945.
37. Mutabazi I., Finn P.L., Gleeson J.T., Goodby J.W. *et al.*, 1992, *Europhys. Lett.*, **19**, 391.
38. Cladis P.E., Finn P.L., Brand H.R., 1995, *Phys. Rev. Lett.*, **75**, 1518.
39. Bodenschatz E., de Bruyn J.R., Guenter A., Cannell D.S., 1991, *Phys. Rev. Lett.*, **67**, 3078.
40. Müller S.C., Coullet P., Walgraef D., 1994, *Chaos*, **4**, 439.

41. Géminard J.-C., Holyst R., Oswald P., 1997, *Phys. Rev. Lett.*, **78**, 1924.
42. Helfrich W., 1978, *J. Phys. (France)*, **39**, 1199.
43. Kosterlitz J.M., Thouless D.J., 1972, *J. Phys. C*, **5**, L124.
44. Kosterlitz J.M., Thouless D.J., 1973, *J. Phys. C*, **6**, 1181.
45. Poniewierski A., Holyst R., 1993, *Phys. Rev. B*, **47**, 9840.
46. Holyst R., 1994, *Phys. Rev., Lett.*, **72**, 4097.

8

Different Applications of Free-Standing Liquid Crystalline Films

8.1 TECHNICAL USE OF FOAMS

Under certain conditions, FSLC (soap) films may form foams, which find a broad area of practical applications (mostly technical and industrial). Foams consist of a liquid phase and distributed in it gas bubbles. If the gas concentration is quite small (generally $\ll 50\%$), bubbles are surrounded by a large amount of a fluid phase, and have a spherical form. In this case, a foam represents the gas emulsion in a liquid. A schematic structure of such monodisperse (the gas bubbles having the same size) emulsion foam is shown in Fig. 8.1(a).

If the gas concentration grows and becomes greater than 50–75%, the number of gas bubbles in such an emulsion also increases, the distances between them diminishes and a polyedric foam may be formed. A schematic structure of a monodisperse polyedric foam is shown in Fig. 8.1(b). Such a foam is the maximal-dense packing of faceted hexagonal gas bubbles divided by liquid walls. It is evident that the latter represents free-standing liquid films. In the zone of the intersection of three such walls the so-called Plateau–Gibbs channels, having a quite complex geometry, are formed.

A real foam practically always contains bubbles of different sizes, i.e. it is polydisperse. The geometrical organisation of such a foam may be very complex. It depends on the ratio between the volumes of the gas and the liquid, on the degree of the foam polydispersity and on the manner of the bubbles arrangement.

Note that the existence of a thermodynamically stable fluid foam is possible only if some of the surfactant is diluted in a liquid phase. Indeed, as we know, the presence of a surfactant stabilises freely suspended films, i.e. the foam walls.

Like any disperse system, a foam may be obtained by means of condensation and dispersation methods. When the condensation technique is used, a foam is obtained

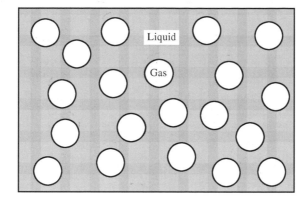

(a)

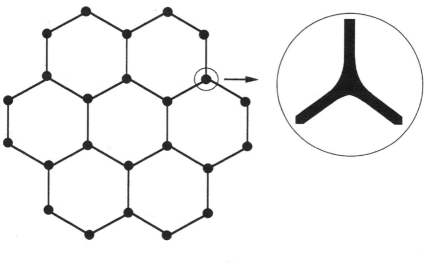

(b)

Fig. 8.1 Schematic organisation of monodisperse foams: (a) gas-in-liquid emulsion; (b) polyedric foam. Solid lines are the liquid walls, solid dots are the Plateau–Gibbs channels (adapted from Reference [2])

in the course of the formation of small gas bubbles in a fluid solution under lowering pressure or increasing temperature (an over-saturated solution is formed). A foam may also be created in the course of a chemical reaction. When the dispersation method is utilised, a foam is obtained by the injection of the gas bubbles into a fluid solution by means of capillary tubes, porous plates, nets or cloth, or by blowing the gas through nets wetted by a solution. Gas dispersion also occurs when a fluid-containing vessel is shaken, when liquid and gas move together in pipes or hoses, or when they are mixed together in special shakers, etc.

The ability of different foams to absorb various gases, liquids and solid particles and to isolate them from the surrounding medium is a property of extreme practical importance. It allows one to use foams for the protection of different objects from dust and for dust absorption, for clearing the objects of solid and liquid pollution, for isolation of aggressive liquids (ammonia, isocyanate, etc.), and for treatment of soils and plants by liquid and gaseous pesticides. This property also makes very promising the application of foams in the textile industry for the coloration of different cloths: the use of foam decreases the water expenditure and increases the homogeneity of the dye distribution. It also makes possible the utilisation of foams for the displacement of oil and natural gas from the soil.

The other important property of foams is their highly developed surface. This property is used in production technology of many nutritive products, which are transformed preliminary into foams. Thus, for example, in the course of the low-temperature drying of milk, coffee, egg powder and potato purée foams are formed.

The mechanical and rheological properties of foams, such as the considerable compressibility, the amortising ability, the high viscosity and the shift stress also find their practical use, for instance, in aviation (the emergency plane landing on foam).

The perfect sound-absorbing ability of some solid foams (e.g. porous polymers) allows their use as noise-isolating materials.

A whole group of methods for the technical quality control of production is based on the foam formation phenomenon itself. For example, observations of foam formation allow one to control both the hermeticity of welded joints of pipes and different vessels and the quality of some products (the foams on beer, champagne, cider, whisky, tea, etc.).

The stability of foams is the essential condition of their practical application. The demands on the degree of foam stability vary for the different above-mentioned applications.

The interested reader may find detailed information on the structure and properties of foams, on their industrial development and practical applications in the monographs in References [1, 2].

8.2 APPLICATIONS OF FOAMS AND SOAP FILMS IN CRYSTALLOGRAPHY AND MATHEMATICS

Initially, the title of this paragraph might look quite absurd. Indeed, what is common between soap films and such abstract sciences as crystallography and mathematics? Nevertheless, a foam has its crystallographic analogue, the crystalline lattice (see, for example, References [3, 4]); and the physico-chemical theory of soap films has its mathematical equivalent, the variation theory of the minimal surfaces (VTMS) (see, for instance, Reference [5]).

The ability of foam bubbles to arrange themselves into hexagonal packing [see Fig. 8.1 (b)] may be used to create models of solid crystalline lattices. Such models were initially proposed by Bregg, and were then used by Lomer and Nye for studying the behaviour of defects (different vacancies and inclusions) and of some other phenomena in solid crystals.

The VTMS is a highly formalised branch of mathematics, so it should use very clear visual models showing the complex topological objects dealt with. Soap films successfully serve as such models, since, in a thermodynamically stable state, they represent objects with a minimal surface (i.e. they have a surface with a minimal area and hence with a minimal surface energy or surface tension).

A simple example of the minimal surface produced by a soap film is a spherical soap bubble. In the absence of gravity and air pressure, any liquid will take a spherical form, since for the given bulk a sphere will have the minimal surface area. Another more general example is a soap film stretched over a closed solid contour of arbitrary form (a wire frame, for instance).

The first to pay attention to the role of soap films in mathematics more than a century ago was J. Plateau (see also Section 1.1). His name was given to an important topic of the VTSM—the Plateau problem. This problem can be formulated as follows: 'Can a minimal surface (soap film) be stretched over any closed boundary contour?'. To answer this question rigorously, one generally should solve a system of non-linear differential equations in partial derivatives. Such equations are very difficult to describe and classify. However, the solution of the Plateau problem may be, in many cases, easily illustrated by stretching soap films over different wire contours. An example of such a contour and two possible soaps film stretching over it (i.e. two solutions of the Plateau problem) are shown in Fig. 8.2.

A soap film stretched over a contour of a quite complex form allows one, for example, to illustrate the notion of a singular point: a point where several film leaves (facets) come together; and to prove the mathematical rule that only two or three film leaves can be united in each singular point. Indeed, let us assume that four film leaves come together at some singular point: the cross-section of such a soap film is shown in Fig. 8.3(a). It is clear that such a configuration is energetically unfavourable, and that this four-leaf point will be immediately transformed into two three-leaf

Fig. 8.2 An example illustration of the Plateau problem with the use of soap films: two different possibilities of stretching a soap film over the same wire contour (adapted from Reference [5])

points [Fig. 8.3(b)]. This structural transformation is accompanied by film deformation, in order to acquire minimal energy (surface area), Fig. 8.3(c). Note that the hexagonal foam shown in Fig. 8.1(b) contains the analogous three-leaf singular points (the cross sections of the Plateau channels).

The problems of minimal surfaces are directly connected with other areas of mathematics, such as topology and multiple-dimensional geometry (the theory of knots). In multiple-dimensional geometry, for example, four-dimensional minimal surfaces (which, for fixed boundary conditions, have a minimal bulk) can be studied. Minimal surfaces also occur in many branches of mechanics and physics, in the theory of field, for instance.

8.3 LYOTROPIC FREE-STANDING FILMS—USE IN BIOLOGY AND MEDICINE

Studies of the physical properties of biological membranes and black soap films (which, as we know, can serve as biomembrane simplified models, e.g. [6–8], see Chapter 7) find a great number of applications in biological and medical sciences. Further on we will consider some examples of such applications.

8.3.1 BIOLOGICAL APPLICATIONS

Investigation of physical properties of black soap films and biomembranes (especially of their elasticity and liquid-crystalline structure) provides a lot of information towards the understanding of the living functions of organic cells.

Before coming to the analysis of the elastic properties of biomembranes, consider the following issue for a model membrane (black film). It is clear that such a membrane has three types of mechanical deformations: stretch, shift and splay. Other

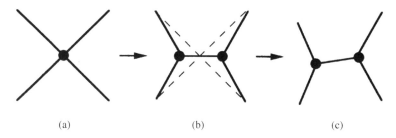

(a)   (b)   (c)

Fig. 8.3 Transformation of one four-leaf singular point into two three-leaf singular points at a soap film surface (adapted from Reference [5]): (a) the cross-section of a four-leaf singular point; (b) the process of its transfer into two three-leaf singular points; (c) the final energetically stable configuration of two three-leaf singular points. Solid lines are the film ribs, solid dots are the singular points

types of deformation typical for mesophases, bend and twist, are prohibited, as is the case for the smectic A liquid crystal (see Chapter 1, Eq. 1.16).

Stretching along the lamellar structure should, in principle, lead to its thinning. Such a deformation, of course, cannot be realised for thermotropic smectics. However, in the case of lyotropic liquid crystals, molecules of which have long supple hydrocarbon tails, this deformation is possible. The same refers to the shift deformation. Indeed, in thermotropic smectics such a deformation transforms smectic A into smectic C, i.e. causes a phase transition, which is why it cannot be realised. In more mobile, lyotropic systems, in contrast, this deformation is possible.

In fact, in the membranes of living cells, all the three above-mentioned types of deformation are observed. The degree of their realisation, however, is different. The biomembrane is very rigid for the stretch and shift deformations and is very soft for the splay deformations. Moreover, the last deformation may be realised both under the action of external forces and spontaneously, if the biomembrane molecules have different chemical composition, or are in contact with different substances which provoke the penetration of some alien particles (such as protein molecules) into the cell membrane (see also Chapter 7, Fig. 7.3).

The free-energy density of the biological membrane, responsible for spontaneous deformation, can be written as follows [9]:

$$F = F_0 + \tfrac{1}{2}K(C_1 + C_2 - C_0)^2 + \tfrac{1}{2}\bar{K}C_1C_2. \tag{8.1}$$

Here F_0 is the free energy density in an undeformed state; K and $\bar{K}$ are, respectively, the splay and the saddle-splay elastic (or rigidity) constants; $C_1 = 1/R_1$ and $C_2 = 1/R_2$ are the main forced curvatures of the biomembrane; R_1 and R_2 are the biomembrane curvature radii in the two mutually perpendicular directions; and C_0 is the spontaneous biomembrane curvature.

The notion of biomembrane elasticity is quite useful for the analysis of geometrical forms of different cells [9]. Let us explain this using the example of the red blood cells, erythrocytes. These cells change their form very easily, which is connected with the necessity to penetrate the thinnest blood capillaries. In the normal state, the erythrocytes have the form of a double-concave discocyte. In some other states they can take the form of a spherocyte (a spherical shape), a stomocyte (a cup-like shape) and an echinocyte (a hedgehog-like shape). The latter form is quite interesting, its surface looks like the packing for eggs.

Changing its shape, the erythrocyte does not alter its surface area. This is connected with the liquid nature of its membrane.

One can think that all the just-mentioned changes of the erythrocyte forms are induced by very complex chemical processes. However, this is not so. These changes are exclusively due to the biomembrane elastic properties—the alteration of the membrane form under the action of pressure. This may be proved by simple calculations.

Indeed, let us add to Eq. (8.1) the term ΔPV (where ΔP is the difference between the external and internal, with respect to the erythrocyte, pressures, and V is the erythrocyte volume). This term gives the free-energy density connected with the pressure. Now one can obtain the various energetically favourable erythrocyte forms by searching for the minimum values of the biomembrane free-energy density (see Eq. 8.1) corresponding to different ΔP.

The results of these calculations—the diametric cross-sections of the three most common erythrocyte forms—are shown in Fig. 8.4. The fourth form—the echinocyte—cannot be obtained by the above-described calculations. This does not mean that some other non-elastic forces play a role in its creation. It is known that the transition of discocyte–echinocyte takes place under the action of certain fatty acids and phospholipids. Probably, in the places where these substances penetrate into the upper layer of the erythrocyte biomembrane, the sign of its spontaneous curvature (C_0) changes from positive (typical for the discocytes) to negative. The latter circumstance leads to the appearance of appendices characteristic of the echinocytes.

The other interesting biological phenomenon, which may also be considered in terms of the cell membrane elasticity, is phagocytosis, the trapping of a substance by a cell [9]. Fig. 8.5 illustrates one of the models of this phenomenon. When an external particle of a substance comes into contact with the surface of the cell membrane, a phase transition occurs locally, at the place of contact: the initially liquid portion of the membrane becomes solid. This leads to a local change of the cell membrane's spontaneous curvature (C_0): the membrane bends around the particle pulling it inside the cell.

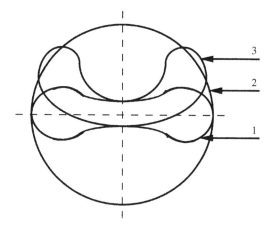

Fig. 8.4 The diametric cross-sections of the main energetically stable forms of erythrocytes (adapted from Reference [9] and reproduced by permission of the author): (1) discocyte; (2) spherocyte; (3) stomocyte

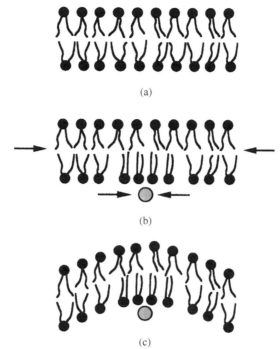

(a)

(b)

(c)

Fig. 8.5 Schematic mechanism of cell phagocytosis (adapted from Reference [9] and reproduced by permission of the author): (a) an undisturbed cell membrane; (b) the approaching external substance particle causing a liquid–solid phase transition in the neighbouring lipid layer section; (c) the membrane bending and drawing the particle inside the cell

The just-described model is extremely simplified. For real biological membranes, one should take into account the electrostatic interaction between the proteins incorporated in the biomembrane, the ion transport and the flexoelectric effect inside the biomembrane [9].

The influence of the magnetic and electric fields upon biological membranes, i.e. the Fredericks transitions (see also Chapter 1), can also be considered in the framework of the elastic theory. To do this, one should add to Eq. (8.1) the terms describing the magnetic or electric energies and to minimise the obtained formula for the free-energy density. The calculations show that the cell membrane will change its form from spherical to elliptical under the action of an external field. In the case of a magnetic field, the cell deformation is small: for a cell of 10^{-5} cm in radius it is only 5×10^{-10} cm for $H = 10^4$ E. The electric field, in contrast, deforms the cell membrane to a greater extent: a cell of 3×10^{-4} cm in radius will be deformed by 3×10^{-5} cm for $E = 30$ V/cm. Such a deformation will influence considerably the cell biological activity. These data may serve for a better understanding of numerous experimental facts of the influence of weak natural electromagnetic (for instance, geomagnetic) fields on living organisms.

Investigation of the rupture process of black films, which can also be described in the framework of the elastic theory (see also Chapter 4), may help to understand better the coagulation of cell membranes, the phenomenon which has essential importance for many biological and biotechnological processes (e.g. [7]).

Finally, the studies of liquid crystalline ordering in biomembranes helps with the solving of some very important questions concerning the origin of life. Indeed, lyotropic liquid crystals formed the membranes of the first cells, the protocells. This process was probably realised as follows (Fig. 8.6). The lipid molecules formed the lamellar phases inside relatively concentrated solutions. This might be realised in dried water basins. Then, with an increasing amount of water, the transition from the lamellar into the micellar hexagonal phase took place. Under further dilution, this phase was transformed into the disordered micellar solution, containing separate protocells.

8.3.2 MEDICAL APPLICATIONS

Studies of the physical properties of black soap films and biomembranes (especially of the structure, thermodynamic stability and phase transitions) may give useful data on the pathologies causing disease in some human organs. This information is extremely important in medicine.

The most reliable data obtained at this moment concern lung disease, the so-called syndrome of respiration dysfunction (SRD) [2, 10]; and the problems connected with the stability of the eye tear layer [7]. Some attempts to relate changes of the liquid crystalline order in the cell-forming structures with the cell cancerous mutation and the ageing process have also been made [7, 9].

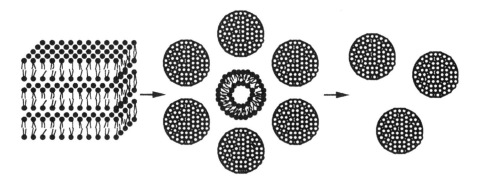

Fig. 8.6 Possible mechanism of creation of protocells. The water content in the system increases from the left Figure to the right one (adapted from Reference [9] and reproduced by permission of the author)

The just-mentioned SRD is directly connected with the stability of the lung alveolar layer. In the normal state, this layer covers the lungs and guarantees alveolar stability and rapid gas exchange. The alveolar layer consists of a thin (of the order of 10–100 nm in thickness) film made of a water solution of some lipids, salts and proteins. This film is covered by a surfactant monolayer mainly formed by phospholipids. The surfactant monolayer ensures a low surface tension between the air and the liquid, thus preventing the rupture of alveoli.

The lack of a phospholipid surfactant in amniotic fluid (the liquid surrounding a human embryo) may cause SRD in new-born children. Indeed, laboratory studies of black, freely suspended, films prepared from amniotic liquid, in dependence upon the phospholipid concentration, c, show that these films are stable when c is higher than some threshold value c_{th}. Thus for $c > c_{th}$, the lung alveoli develop normally, while for $c < c_{th}$, there is the risk of damage of the alveolar layer leading to SRD. In this latter case prevention treatment is needed. Tests on amniotic black films are largely used in prenatal medicine for the early detection of SRD.

The structural studies of some lyotropic FSLC films allow one to explore in detail the molecular organisation in the lung alveolar layer. For instance, very important information has been obtained by studying multi-layer films prepared from alveolar surfactant solutions with the addition of ethanol (47%) and NaCl (0.07 mol/l). The results of this study are represented in Fig. 8.7 as the dependence of the disjoining pressure upon the film thickness: $\Pi(h)$. The $\Pi(h)$ curve shows a stepwise decrease of

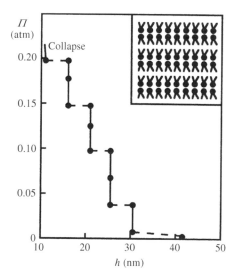

Fig. 8.7 The experimental dependence of the disjoining pressure upon the film thickness for a free-standing film prepared on the basis of lung alveolar phospholipid surfactant. Inset: the assumed molecular structure of such a film (adapted from References [2, 10])

h with a step thickness approximately equal 5.5 nm (structural disjoining pressure, see Section 2.6.3), i.e. the stratification phenomenon takes place. This allows one to assume that the studied film consists of parallel phospholipid molecular bilayers (see the inset to Fig. 8.7). It is possible that such a film may be formed *in vivo* between the membranes of the lung epithelium cells and the phospholipid surfactant alveolar monolayer at excess concentrations of the phospholipids.

An isolated alveolar layer (a free-standing film) should rupture under the action of gravity and capillary forces. However, under physiological conditions, it is stable. A possible mechanism of alveolar film stabilisation is the presence of positive components of the disjoining pressure (e.g. electrostatic repulsive forces) and of the Marangoni effect, which is due to inhomogeneous surfactant distribution in the film.

Effects analogous to those stabilising the alveolar layer may also play a role in the case of the eye tear film. This film consists of three layers: external, intermediate and mucoid. The external layer, which is in contact with the air, is a lipid–mucine film of about 100 nm in thickness. The intermediate one is formed by a water phase containing salts and organic substances. Its mean thickness is about 10 µm. The mucoid layer is situated between the water (intermediate) layer and the eye epithelium. Its thickness is approximately 20 nm. Under normal physiological conditions, this wetting tear film is not stable: if the eye is open, the tear film ruptures in about 30 s.

Now some words about the relation between the structure of FSLC films (biomembranes) and cancer diseases. It is well known that healthy cells are placed in close contact with one another. At the first stage of a cell's malign regeneration, some cavities appear between them; and, in the course of further development of the disease, the contact between the cells is disturbed completely. Cancer cells do not have a regular form and appendices appear at their surfaces. The contamination of the healthy cell always starts by the penetration into it of the malign cell appendix.

In recent years, some attempts have been made to connect the cancer pathologies with the changes in the liquid crystalline order in mesomorphic cellular structures (e.g. inside cell membranes).

These attempts are based on some experimental observations. As has already been mentioned (see Chapter 7, Fig. 7.3) the lamellar structure of the biomembrane lipid bilayer is disturbed by the presence of the protein molecules. Part of these proteins is responsible for inter-cellular contacts: such protein molecules are denoted by the letter R in Fig. 8.8. One can say that the R-proteins glue the cells to one another. In normal cell membranes, the R-protein molecules are distributed homogeneously [Fig. 8.8(a)]. However, in the course of cancerous regeneration, a phase transition takes place: the R-proteins become aggregated in clusters [Fig. 8.8(b)]. Now only the parts of the cell membrane where these protein clusters are present, will be in local contact with other cell membranes. This will lead to the formation of the appendices.

An additional factor playing a role in cancerous cell regeneration is connected with the ordering of the cellular micro-filaments. These are thin threads (of about

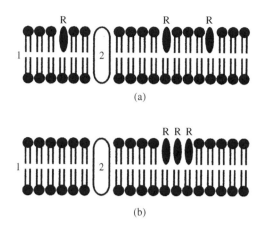

Fig. 8.8 Possible mechanism of cancerous cell regeneration (1, lipid molecules; 2, protein molecules; R, protein molecules responsible for the intercellular contacts): (a) The healthy cell. The R-proteins are distributed homogeneously inside the biomembrane. (b) The malignant cell. The R-proteins form clusters (adapted from Reference [9] and reproduced by permission of the author)

60 Å in diameter) situated in the cell cytoplasm layer, which is in direct contact with the biomembrane. The cellular micro-filaments are similar to the muscle threads. One can say that they serve as muscles for the cell membrane: their contractions change the membrane form. From the structural point of view, the cellular micro-filaments represent a typical smectic liquid crystalline structure.

Experiments show that the filament ordering diminishes considerably for cancer cells. This increases the motional activity of the biomembrane. Inside highly malignant cells, these filaments are tangled into an irregular net. As a consequence, the membranes of such cells are completely out of control.

Macroscopically, the lowering-of-order phase transitions just-described lead to a decrease in the surface tension and the elastic constants and to an increase in the fluidity of the cell membrane.

All the above-said of course does not evoke the fundamental reasons for the malign cellular regeneration. However, it shows that in the course of the detailed study of cancer-related problems, one should necessarily take into account the changes in the FSLC film-like cellular structures.

Finally, investigations of the structure of the cell membranes allows one also to solve some problems concerning the ageing process of living organisms. One of the hypotheses tells us that this process is connected with a decrease of the water fraction in the organism, which leads to a change in the cell membrane mesophase parts to the solid crystalline ones. The latter provokes disturbances in the biomembrane transport functions and leads hence to serious chemical changes inside the cells.

It is evident that to prevent the ageing process, one should increase the amount of water in the organism. However, at the moment, it is not clear how do this.

REFERENCES

1. Bikerman J.J., 1973, *Foams*, Springer, New York.
2. Kruglyakov P.M., Ekserowa D.R., 1990, *Foam and Foam Films*, Moscow, Khimia (in Russian).
3. Geguzin Ya.E., 1985, *Bubbles*, Moscow, Nauka (in Russian).
4. Radchenko I.V., 1965, *Molecular Physics*, Moscow, Nauka (in Russian).
5. Fomenko A.T., 1981, *Khimiya i Zhizn' (Chemistry and Life)*, **12**, 14 (in Russian).
6. Kruglyakov P.M., Rovin Yu.G., 1978, *Physical Chemistry of Black Hydrocarbon Films. Biomolecular Lipid Membranes*, Nauka, Moscow (in Russian).
7. Dukhin S.S., Ruleuv N.N., Dimitrov D.S., 1986, *Coagulation and Dynamics of Thin Films*, Kiev, Naukova Dumka (in Russian).
8. Israelachvili J.N., 1992, *Intermolecular and Surface Forces*, 2nd edn., Academic Press, London.
9. Sonin A.S., 1983, *Liquid Crystals and the Physics of Life*, Znanie, Moscow (in Russian).
10. Exerowa D., Lalchev Z., 1986, *Langmuir*, **2**, 668.

Index